D0214614

Problems for physics students

Problems for physics students

WITH HINTS AND ANSWERS

K. F. RILEY

Lecturer in Physics, Cavendish Laboratory
Fellow of Clare College, Cambridge

CAMBRIDGE UNIVERSITY PRESS

Cambridge

London New York New Rochelle

Melbourne Sydney

Published by the Press Syndicate of the University of Cambridge
The Pitt Building, Trumpington Street, Cambridge CB2 1RP
32 East 57th Street, New York, NY 10022, USA
296 Beaconsfield Parade, Middle Park, Melbourne 3206, Australia

First published 1982

Printed in Great Britain at the University Press, Cambridge

Library of Congress catalogue card number: 82-4575

British Library Cataloguing in Publication Data

Riley, K. F.
Problems for physics students with hints and answers

1. Physics – Problems, exercises, etc.
I. Title
530'.076 QC32

ISBN 0 521 24921 X hard covers
ISBN 0 521 27073 1 paperback

DJ

CONTENTS

PREFACE

This book aims to provide a set of problems which will test a student's understanding of the principles which are usually taught in a tertiary level physics course. In the United Kingdom, for example, the topics covered are those typically met in an A-level physics syllabus. Most of the material is thus what is known as 'classical physics', although all A-level courses contain some 'modern physics', i.e. nuclei, atoms and photons. A few A-level syllabuses, notably the Nuffield one, also contain introductory ideas on the more advanced topics of electronics, entropy and the Schrödinger equation, and questions on these have been included. In countries which adopt a less specialized school curriculum than that found in the UK, the relevant course level is that of junior college or the early years of university.

The difficulty of the questions varies widely, from straightforward application of a single basic idea to quite complex situations involving the use of several ideas at once, some of them not immediately apparent. It is hoped that these more demanding problems will not only stretch the best of pre-university students, but prove of value to those already at university during the earlier parts of their physics courses.

In the first part (sections A–T) of the book I have tried to group problems on similar areas of physics together in one section, the separate sections being alphabetically labelled. In the later part, however, in particular in sections U, V, X and Y, a deliberate attempt has been made to include problems involving ideas from several areas. Within each section similar ideas have in general been grouped together, with those groups which in my view are the more straightforward placed earlier. The questions considered to be the most demanding have been marked with an asterisk (*). It will thus be apparent that the questions under a single letter form a generally increasingly difficult set of problems on one area of physics, and that those with the same number, but different letters, form a roughly uniform set of questions on a variety of topics. A student may therefore take advantage of this structure to meet his or her individual aims or needs.

It seems appropriate here to say a little about the choice of format for the problems posed. Consider the following question and response.

Q. A uniform solid cylinder is set spinning about its axis and is then gently placed, with its axis horizontal, on a rough horizontal table. What happens?

A. At first slipping occurs, but eventually this stops and the cylinder rolls smoothly along the table.

The answer given is, of course, correct. But is it clear that the physics principles underlying the analysis of the situation have been understood? The reply to this latter question has to be 'no'; they may have been, but it is not clear that they have. Were the original question to have asked whether the final linear speed of the cylinder depends upon its mass, its initial angular speed, its radius, the coefficient of friction or the acceleration due to gravity, or what fraction of the initial kinetic energy of the cylinder is dissipated against sliding friction, then answers of no, yes, yes, no and no, or $\frac{2}{3}$, respectively, would be clear indicators that the principles had not only been understood but also correctly applied.

I would not for a minute claim that qualitative discussions of physical problems are not important or difficult; in some cases they are the only discussions that are possible. However, for the reasons illustrated by the previous paragraph, and because it is clearly more practicable when the written word, rather than a face-to-face discussion, is the medium of interaction with the student, I have chosen to make the large majority of problems in this book quantitative. This has the additional advantage that intermediate answers can be provided in a compact form, and so enable the student to locate more readily the part of the analysis in which the reasoning has been at fault, in those cases in which the problem has not been correctly solved. Even though it is not possible in a book like the present one to provide the corresponding answers, since anything offered will inevitably be found by some to be either incomplete or misleading, the importance of descriptive physics is recognized in sections Y and Z, where a significant number of qualitative questions are posed for the student to consider.

Despite the decision to make the questions quantitative, either algebraic or numerical, the mathematical techniques involved are not difficult and should be well within the capabilities of anyone who has studied mathematics beyond O-level; nor are the techniques the main points of the questions, except in the case of the data-handling exercises of section W.

Just as important a purpose of this book as testing, is that of instruction. Not of course in the basic ideas of physics, for which standard texts and teachers are the proper agents, but in the ability to pose to one's self the kind of question which will make it clear which ideas are involved. In doing this by means of hints for the problems, a sometimes difficult balance has to be struck between being so helpful that there is nothing left to the problem, and being so oblique that the hint is merely one more baffling aspect. I hope that in the majority of cases such a balance has been found. The same kind of considerations have applied to the intermediate answers which appear mixed in with the hints. Clearly they do not make the same kind of qualitative suggestions to the student as do the hints, but they should serve to indicate where a calculation has gone off the rails.

The hints and final answers are to be found in separate sections towards the end of the book. A letter H in square brackets at the end of a question indicates that, if needed, a hint or intermediate answer is available for that question or part-question. In some cases answers are given in the questions themselves. Separate listings have been used so as to enable a student requiring help to obtain it without 'accidentally' noticing the final answer.

In order to make the book self-contained for its own purposes, I have included the values of standard constants on the very last page of the book, an alphabetical list of symbols used in the questions, hints and answers (these I have tried to keep in accord with the recommendations of *SI Units, Signs, Symbols and Abbreviations* published by the Association for Science Education), and a list of formulae and relationships such as is used in some A-level courses in the United Kingdom.

It is a pleasure to record my sincere thanks to Heather Cuff, Belinda Powell and Sue Arnold for their patient and careful typing of a difficult text.

My thanks also go to the Cambridge University Tutorial Representatives for permission to base many of the problems on questions set in the Cambridge Colleges' Examination. The suggested answers and hints are of course my own, as they are for the original problems included, and in no way represent solutions officially approved by the Cambridge Colleges. Also my own are all errors and ambiguities, and I would be most grateful to have them brought to my attention.

Finally I wish to place on record my appreciation of the help given by the staff and advisors of the Cambridge University Press with the presentation of this book.

K. F. R.

Cambridge, 1982

A PHYSICAL DIMENSIONS

In this section symbols not explicitly defined have the meanings indicated in the table of symbols and units (p. 171) or in the list of constants given on the last page of the book.

A1 According to Bohr's theory of the hydrogen atom, the ionization energy is $m_e e^4/8\epsilon_0^2 h^2$. Show that this expression has the dimensions of an energy and a value of 13.5 eV. [H]

A2 The Wiedemann–Franz law states that under certain conditions the electrical conductivity σ of a metal is related to its thermal conductivity λ by the equation

$$\frac{\lambda}{\sigma T} = \frac{\pi^2}{3}\left(\frac{k}{e}\right)^2.$$

Show that this equation is dimensionally acceptable, and estimate the thermal conductivity of copper at room temperature, given that its electrical conductivity is $5.6 \times 10^7\,\mathrm{S\,m^{-1}}$. [H]

A3 Examine the following (supposed) equations for dimensional plausibility:

(i) The velocity v of surface waves of wavelength λ on a liquid of density ρ, under the influence of both gravity and surface tension γ;

$$v^2 = \frac{g\lambda}{2\pi} + \frac{2\pi\gamma}{\rho\lambda}.$$

(ii) The energy flux S (the magnitude of the Poynting vector) associated with an electromagnetic wave in a vacuum, the electric field strength of the wave being E, and the associated magnetic flux density B;

$$S = \frac{1}{2}\left[\left(\frac{\epsilon_0}{\mu_0}\right)^{1/2} E^2 + \left(\frac{\mu_0}{\epsilon_0}\right)^{1/2} B^2\right].$$

(iii) The relativistic Schrödinger equation (the Klein–Gordon equation) for a spinless particle, which gives the development of the wave function ψ describing the motion of a π-meson of mass m_π and arbitrary energy;

$$\frac{\partial^2\psi}{\partial x^2}+\frac{\partial^2\psi}{\partial y^2}+\frac{\partial^2\psi}{\partial z^2}=\frac{1}{c^2}\frac{\partial^2\psi}{\partial t^2}+\frac{4\pi^2 m_\pi^2 c^2}{h^2}\psi.$$

The symbols x, y, z and t indicate the usual space and time coordinates.

[H (ii), (iii)]

A4 (i) Assuming that the flow of liquid in a tube becomes turbulent at a critical velocity v which depends upon the viscosity η and density ρ of the liquid, and the radius r of the tube, find how v varies with these quantities.

(ii) If turbulence sets in when the velocity of flow exceeds $4.0\,\mathrm{m\,s^{-1}}$ for water flowing in a tube of radius 5.0 mm, at what flow velocity will it occur for olive oil, which has a density of $9.3\times10^2\,\mathrm{kg\,m^{-3}}$ and a viscosity of $1.0\times10^{-2}\,\mathrm{kg\,m^{-1}\,s^{-1}}$ when it flows in a tube of radius 15 mm? The viscosity of water is $1.0\times10^{-3}\,\mathrm{kg\,m^{-1}\,s^{-1}}$.

A5 The power required by a helicopter when hovering depends only upon the vertical thrust F its blades provide, their length l, and the density ρ of air. By what factor is the power requirement increased when it takes on a load which doubles its weight?

[H]

A6 The radiation emitted per unit time by unit area of a 'black body' at temperature T is σT^4, where σ is the Stefan–Boltzmann constant which has a value of $5.7\times10^{-8}\,\mathrm{W\,m^{-2}\,K^{-4}}$. The constant σ can also be expressed in terms of k, h and c as

$$\sigma=\mu k^\alpha h^\beta c^\gamma,$$

where μ, α, β and γ are dimensionless constants. Determine the numerical value of μ.

[H]

A7 A cobweb of length l, mass m, and under a tension F is contained in a glass case at a temperature T. Because it is struck by air molecules it has random vibrations with energy a few times kT. Determine how the amplitude of these motions can depend upon the quantities mentioned.

A8 The radius R of the fireball of an atomic bomb exploded in an atmosphere of density ρ may be assumed to depend upon ρ, upon the time t after the explosion and upon the energy E released by the bomb.

(i) Check this as far as possible from the following measurements made on an explosion at sea-level:

t/ms	R/m
0.24	19.9
0.66	31.9
1.22	41.0
4.61	67.3
15.0	106.5
53.0	175.0

(ii) Assume the unknown dimensionless constant to have a value close to unity and hence estimate the energy of the explosion. (Ref. G. I. Taylor, *Proc. Roy. Soc.* A, **201**, 175, 1950.)

A9 A new system of units is devised in which unit length is 1 m, but the units of time and mass are so chosen that c and G are both of magnitude unity. What is the new unit of mass in kg? [H]

A10 Elementary particle physicists prefer to work with a system of units in which c and m_p have unit value and h has the value 2π. Find the unit of time in this system. [H]

A11* The periodic time T for a moon of mass m_1 to describe an elliptical orbit of major axis a about a planet of mass m_2 depends on m_1, m_2, a and G.

(i) Determine as far as possible the relationship between these quantities.

(ii) How would the periodic time compare with that T' of another system with masses $2m_1$ and $2m_2$ but the same major axis? [H]

A12* An incompressible fluid flows through a small hole of diameter d in a thin plane sheet. The volume flow rate R depends on d, on the viscosity η and density ρ of the fluid, and on the pressure difference p between the two sides of the sheet.

(i) Find the most general relationship between these quantities.

(ii) Measurement of the flow rate R_1 for a pressure difference p_1 is made using a particular fluid. What can be predicted for a fluid of twice the density and one-third the viscosity? [H (i), (ii)]

A13* (i) The envelope of a jet formed by a liquid escaping from an orifice has a form periodic with distance from the orifice. If it is assumed that the wavelength of this envelope depends only upon the surface tension σ, the density ρ, the pressure p due to the head of liquid behind the orifice, and on the area A of the orifice, obtain by dimensional analysis as much information as possible about the relationship between these quantities.

In a particular experiment with water the following values were obtained:

Head of water/m	0.2	0.4	0.6	0.8	1.0
Wavelength/mm	3.5	4.1	4.8	5.6	6.5

Determine:

(ii) the wavelength at a head of 0.3 m for a liquid whose surface tension is one-third that of water, but has the same density;

(iii) the head needed for a liquid of surface tension one-half and density twice that of water in order to produce a wavelength of 5.0 mm. [H (ii), (iii)]

B LINEAR MECHANICS AND STATICS

B1 An empty cylindrical beaker of mass $m = 100$ g, radius $r = 30$ mm, and negligible wall thickness has its centre of gravity a height $h = 100$ mm above its base.

(i) To what depth x should it be filled with water (density 10^3 kg m^{-3}) so as to make it as stable as possible?

(ii) What is then the height of the centre of gravity of the partially-filled beaker above its base?

(iii) Explain the connection between the two values. [H (i)]

B2 A ladder, 5 m long and of mass M, rests on a rough horizontal floor and against a smooth vertical wall. The maximum distance its foot may be from the wall before slipping occurs is 4 m. When its foot is 3 m from the base of the wall, what is:

(i) The maximum mass that can be placed anywhere on the ladder without causing slipping?

(ii) The maximum distance up the ladder a (point) man of mass $5M$ can safely go? [H]

B3 A light rope carries a mass of 100 kg on one end and is wound around a horizontal cylindrical bar, the coefficient of static friction μ between the two being 0.05.

(i) By considering the equilibrium of a small portion of the rope in contact with the bar, show that the force on the free end needed to support the mass depends exponentially on the number of turns of the rope in contact with the bar.

(ii) Find the minimum number of turns required if the only available counterweight has mass 1 kg. [H (i)]

B4 A particle of mass m carries an electric charge Q and is subject to the combined action of gravity and a uniform horizontal electric field of strength E. It is projected in the vertical plane parallel to the field at a positive angle θ to the horizontal. Show that the horizontal distance it has travelled when it is next level with its starting point will be a maximum if $\tan 2\theta = -mg/EQ$. [H]

B5 An engine of mass M works with a constant tractive force F against a resistance proportional to the square of its speed. The maximum speed it can reach is U. Calculate (i) the time it takes, and (ii) the distance covered, whilst it accelerates from rest to a speed $\frac{1}{2}U$. [H (i), (ii)]

B6 A free-standing wall of height h and thickness t is made of bricks of density ρ_W, and rests on a rough floor. A wind of speed V blows against the wall.

(i) Assuming that the air is stopped when it reaches the wall, show that the wall will topple over if V exceeds $(\rho_W g/\rho_A h)^{1/2} t$ where ρ_A is the density of the air.

(ii) If $\rho_W = 3 \times 10^3 \text{ kg m}^{-3}$, $\rho_A = 1.25 \text{ kg m}^{-3}$, $h = 2$ m and $t = 0.1$ m, find the critical value of V, and determine the minimum value of the coefficient of friction between the floor and the wall for the wall to topple rather than slide. [H (ii)]

B7* Calculate the power theoretically extractable from the wind by a windmill, using the following model.

Let the wind far up- and down-stream of the mill be one-dimensional with speeds V and αV respectively ($0 \leqslant \alpha \leqslant 1$), and let the wind speed at the sails, which sweep out an area A, be v.

(i) By equating the power absorbed by the mill to the rate of loss of kinetic energy of the wind, show that $v/V = \frac{1}{2}(1+\alpha)$.

(ii) Show that the power obtainable is proportional to $A\rho V^3$ where ρ is the density of the air, and that its maximum value is 16/27 of the initial power available in the wind. [H (i), (ii)]

B8 (*a*) A bicycle stands at rest, prevented from falling sideways but able to move forwards or backwards, and with its pedals in their highest and lowest positions. A man crouches beside the bicycle and applies a horizontal force directed towards the back wheel to the lower pedal. (i) Which way does the bicycle move? (ii) In which sense does the chain-wheel rotate (as viewed by the man)? (iii) Which way does the lower pedal move relative to the ground? Explain your answers carefully.

(*b*) A chair and a jug are placed on the platform of a weighing machine and a barrel of beer is then placed on the chair with its tap above the jug. When the tap is open and beer runs into the jug, does the machine register a higher, a lower, or the same reading as before the tap was turned on? [H *a,b*]

B9 A body of mass M moves with kinetic energy E, but without rotation. Because of an internal spring mechanism it divides into two non-rotating rigid bodies of masses αM and $(1-\alpha)M$ which now move in directions making equal angles θ on either side of the original direction of motion. Show that the spring mechanism must have released energy at least as great as $E\tan^2\theta$. [H]

B10 Both stages of a two-stage rocket are propelled by ejecting gas backwards with speed u relative to the rocket. The rocket starts from rest in a field-free region, and its two stages have masses M_1 and M_2 when filled with fuel, and masses m_1 and m_2 when empty. The empty first stage, of mass m_1, is detached before the second stage is fired.

(i) Find the final velocity of the empty second stage.

(ii) Show that it exceeds by $u\ln[M_2(m_1+m_2)/m_2(m_1+M_2)]$ that achieved by a single-stage rocket of mass M_1+M_2 when fully fuelled and mass m_1+m_2 when empty. [H (i)]

B11 Two steel balls of masses M and m are suspended by vertical strings so as to be just in contact with their centres at the same height. The ball of mass M is pulled to one side, keeping its centre in the vertical plane which originally contained the centres. It is released from rest when its height is h above the original position.

(i) Show that, whatever the value of m, the second ball cannot rise to a height above its equilibrium position greater than $4h$.

(ii) A similarly-suspended third steel ball of mass μ is now added (just touching that of mass m when at rest) and the ball of mass M is again drawn aside and released. Show that the kinetic energy transferred to the third ball is a maximum if m is chosen to have the value $(M\mu)^{1/2}$. Assume that all collisions are elastic. [H (i), (ii)]

B12* An aeroplane flies at constant speed V relative to the air and completes a level circular course in time T on a windless day. Show that if a steady wind of speed kV blows in a fixed horizontal direction, then the time for the course is increased by approximately $\frac{3}{4}k^2T$, provided that $k \ll 1$. [H]

B13* Two small smooth blocks A and B with equal masses are free to slide on a frozen lake. They are joined by an elastic rope of length $2^{1/2}L$, of negligible mass, and having the property that it stretches very little when it becomes taut. At time $t=0$, A is at rest at $x=y=0$ and B is at $x=L, y=0$ moving in the positive y-direction with speed V. What are the positions and velocities of A and B at times (i) $t=2L/V$, and (ii) $t=100L/V$? [H]

C CIRCULAR AND ROTATIONAL MOTION

C1 A wheel, set with its axis vertical, has on one of its horizontal spokes a point mass mounted so that it can slide freely. The mass is connected to the centre of the wheel by a light spring. When the wheel is turned at angular frequency ω the spring is f times its unstretched length. Find the general relationship between ω and f if $f = f_0$ when $\omega = \omega_0$. [H]

C2 A test-tube 100 mm long is filled with water and spun (in a horizontal plane) in a centrifuge at 300 rev s^{-1}. What is the hydrostatic pressure on the outer end of the tube if the inner end is at a distance of 50 mm from the axis of rotation? [H]

C3 A body suspended at rest from a fixed point by a light elastic string of unstretched length l_0 produces an extension l_1. Show that, if it then moves in a horizontal circular path (as a conical pendulum), the period of revolution is $2\pi[(l_0 \cos\theta + l_1)/g]^{1/2}$ where θ is the angle the string makes with the vertical. [H]

C4* (i) Show that, if the inclination θ to the vertical of the inextensible string of a conical pendulum is small, the period of the pendulum is independent of θ.

A small ring is threaded on the string and held so that it provides a point of oscillation. The conical pendulum is set in motion with angular velocity ω_0 and angular momentum J, and its length is then slowly shortened by sliding the ring downwards.

(ii) Find expressions for the pendulum's kinetic and potential energies in terms of J and its angular velocity ω.

(iii) Determine directly the work done in sliding the ring down the string.

Assume that the inclination of the string to the vertical is always small.

[H (i), (ii)]

C5* A cylindrical vessel of height 0.12 m and radius 0.06 m is two-thirds filled with a liquid. The vessel is rotated with constant angular velocity ω about its axis, which is vertical.

(i) Show that if surface tension is neglected the free surface of the liquid is part of a paraboloid of revolution ($z = cr^2$).

(ii) Estimate the greatest angular velocity of rotation for which the liquid does not spill over the edge of the vessel. [H (i), (ii)]

C6 One end of a uniform rod is placed at the edge of a very rough table and the rod is released from rest in an almost vertical position. As it falls away from the table it loses contact with the table when the reaction along the rod becomes zero. Show that this happens when the rod is inclined at an angle $\cos^{-1}(3/5)$ to the vertical. [H]

C7 A horizontal turntable in the form of a uniform disc of mass 150 kg is mounted on a light, frictionless vertical axis at its centre. Two men each of mass 75 kg stand at opposite ends of a diameter, and they and the turntable are at rest. The men then move round the table in the same direction and at the same constant speed. Calculate the angle they have turned through in space when they have made one complete circuit of the table. [H]

C8 A uniform laminar disc of radius a and mass M is held in frictionless bearings with its axis horizontal. The disc is at rest when a fly, also of mass M and moving in a horizontal line in the plane of the disc, lands without significant slipping on the lowest point of the perimeter of the disc at time $t = 0$. In the subsequent motion the disc turns through half of a revolution (carrying the fly with it) before coming to rest.

(i) Find the angular velocity, ω_0, of the system just after the fly has landed.

(ii) Show that the fly is level with the axis of the disc at time $t = T$, where

$$T = \frac{1}{\omega_0} \int_0^{\frac{1}{2}\pi} \sec\left(\tfrac{1}{2}\theta\right) \, d\theta .$$

[H (i)]

C9 Water flows at a horizontal velocity V into the top buckets of a mill-wheel of radius r and is tipped out when the buckets reach the bottom.

(i) If the mass flow rate of the water is M, find the torque on the wheel when it has angular velocity ω.

(ii) Show that the power generated is a maximum when $V = 2\omega r$.

(iii) Calculate the theoretical efficiency of the system for this condition.

Assume that the water in the buckets is uniformly and thinly spread round the wheel, and that the centre of gravity of a semicircle of radius r is $2r/\pi$ from the centre of the corresponding circle. [H (i)]

C10 (i) Determine the moment of inertia of a spherical body of mass m and radius a which has a density distribution inversely proportional to the distance from the centre of the sphere.

(ii) Supposing that the Earth has a density distribution of this form and that due to the tides caused by the Earth's rotation about its own axis 3×10^{19} J are dissipated as heat per annum, calculate the change in rotational period each year. Mass of Earth $= 5 \times 10^{24}$ kg. Radius of Earth $= 6 \times 10^{6}$ m. [H (i), (ii)]

C11 (*a*) A uniform rod of mass M and length l pivoted freely at one end, initially hangs at rest. The rod is struck by a horizontal blow of impulse P at a distance y below the pivot. Find how R the impulsive reaction at the pivot, resulting from the blow, varies with y. The moment of inertia of the rod about an axis through one end perpendicular to its length is $\frac{1}{3}Ml^2$.

(*b*) If the mass in a cricket-bat blade of length l can be taken as distributed approximately proportional to $1 + (x/l)$, where x is the distance from the bat handle, how far down the blade should a batsman aim to have the bat strike the ball, assuming he holds the handle close to the blade? [H *a*, *b*]

C12* A particle of mass m is attached to a pin O on a smooth table by a light string of length $2R$. The centre of the string is initially held at O and the particle set in uniform circular motion (of radius R) about O with angular velocity ω_0. The centre of the string is then released.

(i) Find the subsequent impulse on the pin.

(ii) Find the final kinetic energy of the particle and reconcile your result with energy conservation. [H (i)]

C13* A solid garden roller runs down a hill of height h and out onto a horizontal road. It then crashes into a smooth rigid vertical wall, the impact being (somewhat improbably) perfectly elastic. What is its speed u when it stops skidding and begins to roll again, if it is still on the horizontal road when this happens? [H]

D GRAVITATION AND CIRCULAR ORBITS

$G = 6.7 \times 10^{-11}$ N m^2 kg^{-2},

$g = 9.8$ m s^{-2},

Earth's radius $R_E = 6.4 \times 10^6$ m,

Earth-Sun distance $R_{ES} = 1.5 \times 10^{11}$ m.

D1 From the fact that the Moon orbits the Earth in approximately 28 days, estimate the distance of the Moon from the Earth's centre. [H]

D2 A double star consists of two stars, each of the same mass as the Sun, which are observed to rotate around each other in one week. Find their separation d. [H]

D3 Two geostationary satellites are used to communicate between points on diagonally opposite sides of the Earth. Calculate the delays you would expect due to the finite velocity of light. [H]

D4 Charged dust particles, each carrying one elementary charge, and all of the same mass, form a stable uniform cloud. Find the mass of each particle. [H]

D5 An Earth satellite of mass 20 kg is in a circular orbit at a height small compared to the radius of the Earth.

(i) If air resistance causes its total energy to decrease by 10 kJ per revolution, what is the fractional change in its speed per revolution?

(ii) Does the speed increase or decrease? [H (i)]

D6 A star, around which a small planet is in circular orbit, radiates energy isotropically and consequently its mass decreases slowly from M_1 to M_2. Deduce the ratios of the new to old values for the planet of (i) the orbital radius, (ii) the frequency of rotation, and (iii) the total energy, in terms of $\lambda = M_2/M_1$. Make clear the signs of the changes. [H]

D7 A particle moves in a field of force that varies inversely as the cube of the distance between the particle and the origin and is directed towards the origin. Find its total energy and angular momentum when it moves in a circular orbit of radius r and comment on the result. [H]

D8 Two similar spacecraft are launched from a space station orbiting the Earth by rockets which burn for only a few minutes. Spacecraft A is launched so that it just escapes from the solar system. Spacecraft B is launched in such a way that it falls into the centre of the Sun. Show that spacecraft B requires a more powerful rocket to launch it than does spacecraft A. Assume that the Earth moves in a circular orbit round the Sun, and ignore both the velocity of the space station relative to the Earth and the Earth's gravitational field. [H]

D9 Two solid copper spheres of radii 10 mm and 20 mm are released from rest in free space, their centres being 0.2 m apart. Find the relative velocity with which they eventually collide. Density of copper $= 8.9 \times 10^3$ kg m^{-3}. [H]

D10 (*If no use is made of the hints.) Two equal spherically symmetrical stars, A and B, each of mass M and radius r have their centres $6r$ apart. Make a sketch showing the lines on which the gravitational potential takes the values (i) $-10GM/11r$, (ii) $-2GM/3r$, (iii) $-GM/3r$. Accurate plotting is not required, but the general shapes of the lines need to be clear. (iv) What is the minimum velocity with which gas can be emitted from the surface of B and still be captured by A? Ignore any effects due to the motion of the stars. [H (iv)]

D11 Light may be considered as a stream of particles (photons) each having a mass $h/\lambda c$ and an energy hc/λ, where λ is the wavelength of the light. Light of wavelength 500 nm is emitted from the surface of the Sun and received on Earth slightly shifted in wavelength.

(i) From the following data estimate the fractional shift; radius of Sun $= 7 \times 10^8$ m, mass of Sun $= 2 \times 10^{30}$ kg.

(ii) Is it permissible to neglect the effect of the Earth's gravitational field? [H (i)]

D12 According to Einstein's theory of general relativity a small correction $6GMmv^2/r^2c^2$ should be added to the Newtonian gravitational force between a sun of mass M and a planet of mass m moving with speed v in a circular orbit of radius r around the sun.

(i) Show that this correction decreases the period of motion of the planet by a factor $1-3GM/c^2r$.

(ii) Estimate the advance of the planetary motion of Mercury in a century due to this correction.

The mass of the Sun is 2.0×10^{30} kg and the radius of Mercury's orbit, which is assumed to be circular, is 5.8×10^{10} m. [H (i), (ii)]

D13* Two stars A and X with masses m_A and m_X and separation d revolve in circles around their common centre of gravity under the influence of (Newtonian) gravity.

(i) Find the velocity of A and the angular velocity ω.

(ii) An observer views the system from a very large distance at an angle θ to the normal to the plane of their orbits. He can recognise A as a star of a type that has mass 30 units. Star X, which produces X-rays, must be either a neutron star or a black hole, which depends upon whether its mass is less than or more than 2 units respectively. He measures the line-of-sight component of A's velocity and finds that it is of the form $K \sin(\omega t + \epsilon)$. Show that

$$\frac{m_X^3}{(m_A + m_X)^2} \sin^3 \theta = \frac{|K|^3}{\omega G}.$$

(iii) His measurements give $|K|^3/\omega G = 1/250$ units. If all values of $\cos\theta$ are equally likely, find the probability that X is a black hole. [H; H (iii)]

E SIMPLE HARMONIC MOTION

E1 (*a*) A small mass suspended from a light helical spring is pulled down 20 mm from its equilibrium position and then released from rest. It first returns to the position of release after 2.0 s. (i) What velocity does it then have? (ii) If the motion is described by the equation $x = a \sin(\omega t + \epsilon)$, where x denotes distance from the equilibrium position and t the time elapsed since the initial release, determine values for a, ω and ϵ.

(*b*) A pure crystalline solid may be represented as an array of regularly-spaced atoms, each of mass 4×10^{-26} kg, and all of which vibrate with simple harmonic motion of frequency 3×10^{14} Hz. If the thermal energy of a single atom is 6×10^{-21} J, find (i) the amplitude of vibration of each atom, and (ii) the maximum speed of each atom during its vibrations. [H *b*(i)]

E2 A particle executes s.h.m. about $x = 0$ with angular frequency $\omega = 1$ rad s^{-1}. Determine the amplitude of the motion if measurements on the particle's motion give the following results. The two sets of measurements refer to different motions.

(i) At the same time, $x = 3.0$ and $\mathrm{d}x/\mathrm{d}t = -4.0$.

(ii) $\mathrm{d}x/\mathrm{d}t$ has the value 4.0, and one second later has the value -3.0. [H (ii)]

E3 A particle of mass m moves in a one-dimensional potential given by $V(x) = \frac{1}{2}kx^2$. In a small region of length l around the origin ($x = 0$) the mass experiences a frictional retarding force of constant magnitude f. If the mass starts from rest at the point $x = x_0$ ($x_0 \gg l$), estimate the total time required for the mass to come to rest near the origin. [H]

E4 A body of mass 2 kg rests on a horizontal platform which moves vertically in such a way that at time t its height above its mean position, measured in metres, is given by $y = 0.1 \sin \omega t$. The angular frequency ω is slowly increased, the amplitude being kept constant. At an angular frequency ω_1 the body just begins to lose contact with the platform. Find (i) the value of ω_1, (ii) the maximum force exerted by the platform on the body when $\omega = \omega_1$, and (iii) the maximum height reached by the body if the angular frequency is suddenly doubled to $2\omega_1$ when the platform is in its lowest position.

[H (i), (ii), (iii)]

E5 (*a*) A weighted test-tube floats vertically to a depth l in a container of liquid. It is depressed a little from its equilibrium position and then released. Find the period of subsequent oscillation, assuming that the tube has a uniform cross-section.

(*b*) The density of a liquid increases linearly with depth. Its density at the surface is ρ_0 and at depth D is $2\rho_0$. A small rigid sphere of density $2\rho_0$ is released from rest at a depth $\frac{1}{2}D$.

(i) Neglecting friction and any variation in fluid pressure due to the motion, show that the sphere will execute vertical s.h.m.

(ii) What are the amplitude and period of this oscillation? [H *b*]

E6 A particle of mass m moves in a potential $V(x)$. Show by means of a qualitative sketch-graph how you would expect the frequencies of the oscillations to vary with their amplitudes A for (i) $V(x) = \frac{1}{2}kx^2/(1 + b|x|)$ with $b > 0$, (ii) $V(x) = k(1 - \cos x)$, and (iii) $V(x) = k(\cosh x - 1)$. [H]

E7 A pendulum clock is mounted on a horizontal gramophone turntable with the point of suspension of the pendulum above the centre of the turntable. When the turntable is at rest the pendulum oscillates with small amplitude and period 1 s.

(i) What will the period be when the turntable rotates at $33\frac{1}{3}$ r.p.m.?

(ii) Determine the lowest turntable speed at which the pendulum ceases to oscillate.

The pendulum always swings in the same plane with respect to the clock.

[H (i)]

E8 A uniform circular disc is supported in a horizontal plane by a number of light vertical threads attached at points equally spaced around its circumference, and each $l = 1$ m in length. Find the period of small torsional oscillations of the disc in its own plane. The moment of inertia of a uniform disc of radius a and mass M about an axis through its centre and perpendicular to its plane is $\frac{1}{2}Ma^2$. [H]

E9 A flywheel is carried by a light cylindrical axle of radius 20 mm. The ends of the axle are mounted horizontally in frictionless bearings and a light thin string, carrying a mass of 2 kg at its free end, is attached to and wound round the axle. When the system is released from rest the mass falls through a distance of 5 m in a time of 2 s. The flywheel is now suspended, with its axle vertical, from the end of a torsion wire of constant 4×10^{-4} N m rad^{-1}, twisted through a small angle, and released from rest. How long will it be before it comes to rest again? [H]

E10 Two small pegs are placed 1 m apart, one vertically above the other. To the upper one is attached a light inextensible string of length 2 m carrying a compact mass at its lower end. At time $t = 0$ the pendulum so formed is released from rest with θ, the angle between the vertical and the lower end of the string, small and equal to θ_0, and with the string taut. Sketch the time-variation of θ. Neglect frictional losses. [H]

E11 (*a*) A body is suspended by a spring inside a closed container which is full of a light viscous liquid. If the container is forced to oscillate with s.h.m. so that its vertical displacement at time t is $z = A \cos \omega t$ the mass will also oscillate, its vertical displacement (with respect to the laboratory) being $x = B \cos(\omega t + \phi)$. It can be shown that when ω is equal to ω_0, the frequency of undamped oscillations when the container is at rest,

$$\frac{B}{A} = \left[1 + \left(\frac{k}{b\omega_0}\right)^2\right]^{1/2}, \qquad (\dagger)$$

where k is the spring constant and b is the drag coefficient for the body (drag force per unit velocity through the liquid). One particular body is found to extend the spring by 0.1 m when hung on it, and to reach a terminal velocity of 2 m s^{-1} when falling freely through the liquid. Find the value of B/A for this body at frequency ω_0.

(*b*) With the oscillations at a very high frequency ($\omega \gg \omega_0$), the motion of the container is arrested when it is at its highest position. Sketch the motion of the body during a time interval extending to a few natural periods on either side of the moment of arrest. Assume that the damping is light.

(*c*)* Prove the relationship (†).

(*d*) Use the equation of motion constructed to answer (*c*) to further annotate the sketch in (*b*) and confirm (or correct!) such quantities as the size of the initial amplitude and the phase of the mass's motion at the moment of arrest. [H *c*, *d*]

E12 A body of mass M is connected by two similar, light, stretched springs to two fixed points, and moves along the line joining them. A second body of mass m, initially at rest, is added to the first and sticks to it. By what factor does the energy of the oscillation change if the second body is added (i) when the first body is at its position of maximum velocity, and (ii) when the first body is at an extreme position of its oscillation? [H (i)]

E13 (*a*) Particles A and B, each of mass M, are joined to fixed points, C and D respectively, by light stretched springs of spring constant k, and to each other by a third light stretched spring of constant $\frac{3}{2}k$. All components are supported on a smooth horizontal table and C, A, B, D, in that order, lie on a horizontal line along which all motions take place. The particles are released from rest after both have been moved small distances x_0 from their equilibrium positions (i) in the same direction, and (ii) in opposite directions. Show that in each case A and B have the same equation of motion as each other, and that the ratio of the frequencies of the subsequent oscillations in the two cases is 1 to 2.

(*b*)* Particle A is now moved a distance x_0 and B held fixed before the system is released from rest at $t = 0$. Construct a rough graph of the subsequent displacement of A up to a time $t = 2\pi(M/k)^{1/2}$.

(*c*)* What happens to B? [H *b, c*]

F WAVES

Take the speed of sound in still air as $350 \, m \, s^{-1}$.

F1 A cork floats on the surface of a pond across which a sinusoidal wave-train of wavelength 10 m and amplitude 0.10 m is travelling. The velocity v of waves of wavelength λ on a liquid surface is given by

$$v^2 = \frac{g\lambda}{2\pi} + \frac{2\pi\gamma}{\lambda\rho},$$

where ρ is the density and γ the surface tension of the liquid, which for water has the value $7 \times 10^{-2} \, N \, m^{-1}$. Find the maximum speed of the cork. [H]

F2 (*a*) A long spring, which has a linear density of $2 \, kg \, m^{-1}$ when unstretched and increases by 1% of its unstretched length for each 10 N of tension in it, is subjected to a tension of 200 N. Find the velocity of (i) transverse waves, and (ii) longitudinal waves in the spring.

(*b*) A 'slinky' spring has many turns of large radius, is made of thin wire, is easily deformed, and may be extended to many times its unstretched length. A transverse pulse travels the length L of an extended slinky spring of negligible unstretched length in time T. If the same spring is now extended to λL, what would be the new travel time? [H *a, b*]

F3 (*a*) A viola string 0.50 m long is tuned to A, a frequency of 440 Hz. (i) By how much must the string be shortened (by fingering) to raise its frequency to 550 Hz? (ii) If it goes out of tune and vibrates at 435.6 Hz, by how much and in which sense must the tension in the string be changed to retune the viola?

(*b*) A piano string 1.5 m long is made of steel of density $7.7 \times 10^3 \, kg \, m^{-3}$ and Young modulus $2.2 \times 10^{11} \, N \, m^{-2}$. The tension in it produces an elastic strain of 1%. What is the fundamental frequency of the string? [H *b*]

F4 (*a*) A vibrating tuning fork held over a narrow tube containing a variable amount of water is found to give resonances when the lengths of the air column are 0.359 m and 1.053 m. In a separate experiment the tuning fork gives beats of 1.5 Hz when sounded together with a standard source of frequency 250.0 Hz the frequency of the beats increasing when a very small piece of Plasticine is attached to one of the prongs of the fork. Find the speed of sound in the air of the tube.

(*b*) A 'dust tube' 1.5 m long is sealed at one end and has the diaphragm of a variable frequency sound generator at the other. Both ends can be taken as nodes. Calculate the frequency of the generator at which the tube shows five antinodes (i) if it contains air, and (ii) if it contains helium. [H *a, b*]

F5 Two sources of sinusoidal sound waves of equal amplitudes are placed 7 m apart. One oscillates at 500 Hz and the other at 1.000 kHz, and their phases are such that at certain times both outputs pass through zero together.

(i) Sketch how the resultant disturbance at a point midway between the sources varies with time.

(ii) How far from the midway point is the closest position at which a maximum resultant disturbance occurs? [H (i), (ii)]

F6 A man walks parallel to and at a distance D from a fence whose regular vertical slats are a distance a apart. What does he hear as the echo from the fence of each footfall? [H]

F7 A radio receiver receives simultaneously two signals from a transmitter 500 km away, one by a path along the surface of the Earth, and one by reflection from a portion of the Heaviside layer situated at a height of 200 km between the transmitter and receiver. The layer acts as a perfect horizontal reflector. When the frequency of the transmitted waves is 100 MHz it is observed that the combined signal strength received varies from maximum to minimum and back to maximum eight times per minute. With what (slow) vertical speed is the Heaviside layer moving? [H]

F8 (*a*) The concrete surface of a straight road has a regular pattern of ridges 5 cm apart running East–West across its width. The tyres of a car travelling North at 25 m s^{-1} on a windless day produce a humming sound on the road. What is the frequency heard by (i) the driver, (ii) a man standing at the roadside, and (iii) a cyclist riding South at 5 m s^{-1}?

What Doppler effect would be experienced in cases (*b*)–(*d*)?

(*b*) An astronomer measuring the wavelength of a spectral line which is at 500 nm in the laboratory, if the line is (i) from a star moving away from the Earth at 50 km s^{-1}, and (ii) from the visible extremities of the solar equator. Sunspots cross the face of the Sun in about 12 days and the Sun's diameter subtends about $\frac{1}{2}^{\circ}$ at the Earth's surface.

(*c*) A boy listening to a distant fairground organ while sitting on a moving roundabout. The roundabout has a diameter of 15 m and rotates 20 times per minute.

(*d*) A man listening to music on a radio in a car moving at 60 km h^{-1} towards the transmitter when the radio is operating on a frequency of 1 MHz and the 'music' is a tone of 256 Hz. [H *b*(ii)]

F9 A car travels at 72 km h^{-1} towards a stationary source of sound waves of frequency 1 kHz. Waves reflected from the car return to the source and are used to produce beats with the original waves. Find the frequency of the beats. [H]

F10 A stationary source emits sound waves of constant frequency f_0. For an observer moving with constant velocity the apparent frequency f of the waves is given as a function of time t by:

t/s	0.0	30.0	60.0	70.0	75.0	80.0	90.0	120.0	150.0
f/Hz	210.4	210.4	210.0	208.0	199.5	185.8	182.1	181.6	181.6

Estimate values for (i) f_0, (ii) the speed of the observer, and (iii) the time at which the observer is closest to the source. [H (i), (iii)]

F11 When light of unit intensity from an unpolarized source is incident on a sheet of polaroid it is found that it emerges plane-polarized with an intensity which is 32% of the incident intensity. A second identical sheet of polaroid is used to absorb completely the light transmitted by the first polaroid. A third identical sheet is placed between these 'crossed polaroids' and initially orientated so that it transmits no light. If it is then rotated in its own plane through an angle θ, what is the intensity of light transmitted through the whole system? [H]

F12* A sound baffle is made by joining two vertical boards at right angles so that the section in a horizontal plane consists of two equal sides OA and OB of an isosceles right-angled triangle. Sound waves from a very distant source S (and of wavelength $\lambda \ll OA = L$) are incident horizontally on the baffle parallel to the direction of the internal bisector of the angle AOB. Show that in the region SAB

lines of (displacement) nodes are formed perpendicular to SO and determine their distances y in front of the line AB. [H]

F13* The transverse displacement (in cm) of a taut string carrying a sinusoidal wave is measured at points along its length close compared to the wavelength of the wave, and at time intervals small compared with the period of the wave. The results are:

	Distance x along the string/m		
Time/s	6	7	8
3	a	5.68	b
4	8.08	10.40	12.62
5	c	14.71	d

Determine the values a, b, c, d, the speed and direction of the wave, and the amplitude of the wave. [H]

G GEOMETRICAL OPTICS

Unless otherwise instructed take the refractive index of glass $=\frac{3}{2}$, of water $=\frac{4}{3}$. Prism formula at minimum deviation, $n=\sin\frac{1}{2}(A+D)/\sin\frac{1}{2}A$. Two thin lenses in contact have an effective focal length $f_1f_2/(f_1+f_2)$.

G1 (i) A small disc lies centrally at the bottom of a cylindrical beaker of height 150 mm and is illuminated from below. A converging lens of focal length 100 mm rests on the rim of the beaker and forms an image of the disc 6 mm in diameter. Where is the image located?

(ii) The beaker is now filled with water to a height of h mm. Determine the size of the disc's image as h varies from 0 to just less than 150. [H (ii)]

G2 Show that the condition for constructive interference between light reflected from the upper and lower surfaces of a thin uniform film of thickness t and refractive index n is

$$2t(n^2-\sin^2\theta)^{1/2}=(m-\tfrac{1}{2})\lambda,$$

where θ is the angle that the incident light of wavelength λ makes with the normal to the film, and m is a positive integer. [H]

G3 An electro-optical device is used to modulate the intensity of a continuous light beam. The transmission of the device is a square-wave function of time with alternate equal periods of full and zero transmission of the incident light. The light, after reflection at a distant plane mirror, passes back through the device and its intensity is then measured. An extinction is obtained at a modulating frequency of 6 MHz, and, if the frequency is increased, the next one occurs at 10 MHz. With the frequency set at 6 MHz a glass sheet is inserted between the device and the mirror, and it is found that the extinction can be recovered by decreasing the frequency by 1 kHz. Find the thickness of the glass sheet. [H]

G4 (*a*) The internal and external diameters of a glass thermometer tube are

1 mm and 3 mm. When the tube is viewed from the side what is the apparent internal diameter (i) approximately, and (ii)* more accurately?

(*b*) The same thermometer hangs vertically in water at the centre of a cylindrical glass beaker 100 mm in diameter. What are the approximate apparent diameters when the thermometer is viewed from a long way off through the walls of the beaker? [H *a*(ii), *b*]

G5 The focal lengths of the collimator lens and telescope objective of a prism spectrometer in normal adjustment are 250 mm and 200 mm respectively. The collimator slit has a width of 1.4 mm and the refracting angle of the prism is 60°. The refractive indices of the prism glass for two neighbouring lines in a spectrum are 1.498 and 1.502. Do the two lines seen through the telescope overlap? [H]

G6 (i) Parallel light falls, at a small angle θ to the normal, on one of the faces of a prism of small refracting angle A. Show that, independent of the value of θ (provided it is small), the light is deviated through an angle $(n-1)A$, where n is the refractive index of the prism glass.

(ii) After passing through such a prism the light falls on a similar prism of angle A' and refractive index n'. Show that by a suitable choice of A and A' (n and n' being given) it can be arranged that lights of two different wavelengths λ_1 and λ_2 can both be deviated through the same angle, say 5°. Assume the following data: for λ_1, $n_1 = 1.504$, $n_1' = 1.752$ and for λ_2, $n_2 = 1.496$, $n_2' = 1.748$. [H (i)]

G7 An astronomical telescope whose objective and eyepiece are simple achromatic lenses with focal lengths 500 mm and 10 mm is focussed to view the Moon in normal adjustment.

(i) How far would the eyepiece tube have to be moved in order to project an image of the Moon onto a plate 50 mm behind the eyepiece lens?

(ii) What will be the size of the image? Take the angular diameter of the Moon as 0.50°.

G8 The eyepiece of a compound microscope has a focal length of 20 mm. The objective, of diameter 8 mm, has a focal length of 2 mm. The lenses are fixed 175 mm apart and the final image is formed 250 mm from the eyepiece, at the least distance of distinct vision. Determine (i) the distance of the object from the objective lens, (ii) the overall magnifying power (angular magnification), and (iii) the diameter of the exit pupil (eye-ring). (iv) Over what range of object distance will an observer with normal vision be able to keep the magnified image in focus? [H; H (iv)]

G9 You are provided with two converging lenses of focal lengths 30 mm and 40 mm and a diverging lens of focal length 45 mm. How can these be best arranged to form (i) an astronomical telescope, and (ii) a compound microscope of total length 150 mm? Find the magnification in each case in normal adjustment.

G10 (i) Show that the size of the image of a distant object formed by a simple converging lens is proportional to the focal length of the lens.

(ii) A simple telephoto lens enables the size of a camera image to be increased without increasing the length of the camera and consists of a combination of a converging objective lens and a diverging lens placed some way behind it. Draw a ray diagram that shows how the effective focal length f' of the combination can be greater than the objective–film distance L.

(iii) If the focal length of the objective is 76 mm and that of the diverging lens 25 mm, with a 60 mm space between them, show that the image formed is approximately twice as large as could be obtained by a single converging lens placed at the objective position and forming an image in the same place. What is the approximate value of L? [H (iii)]

G11* A generalized version of Snell's law can be stated as follows (and proved from Fermat's theorem). 'For a light ray in an xy-plane in which the value of the refractive index n depends only on y, and not on x, $n \sin \phi$ is constant along the ray, where ϕ is the angle between the tangent to the ray and the y-axis.'

(i) A man stands on a long plane concrete runway above which a uniform vertical temperature gradient results in a uniform gradient in the refractive index of the air, $n(x, y) = n_0(1 + \alpha y)$, where $|\alpha| = 1.5 \times 10^{-6}\,\mathrm{m}^{-1}$. As a result he cannot see the runway ($y = 0$) beyond a certain distance d. If his eyes are 1.7 m above the runway, find the value of d.

(ii) Does the temperature rise or fall with increasing height? [H (i)]

G12 (*a*) The imaging properties of a system consisting of a thin convex lens and a plane mirror are similar to those of a concave mirror. One particular system consists of a lens of focal length $f = 0.3$ m placed a distance $L = 0.2$ m from a plane mirror. By considering the special situation where object and image coincide, show by means of an annotated ray diagram how far (s) from the lens a concave mirror having the same imaging properties as the lens–mirror system should be placed.

(b)* Show that if the system is to behave like a concave mirror for general positions of the object and final image (take both on the side of the lens remote from the plane mirror) then s must have the value $fL/(f-L)$ and that the focal length of the equivalent concave mirror is $f^2/2(f-L)$. [H b]

G13* (a) A truncated isosceles triangular glass prism stands, as shown in the diagram, with its base in water. (i) Show that if a ray of light, parallel to the

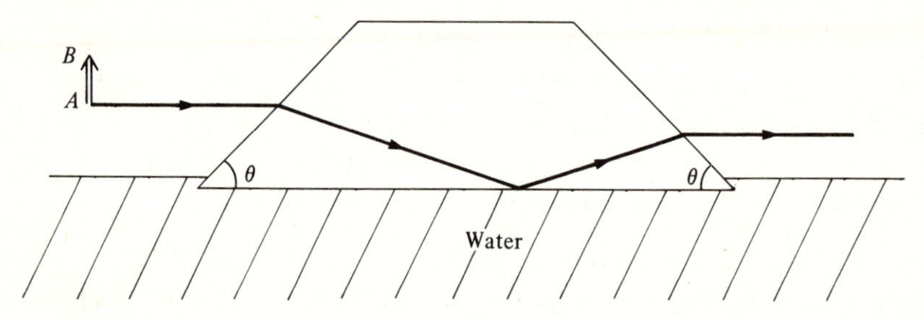

base and lying in the plane of the diagram is incident upon the prism, then it will be totally reflected at the base, provided that the angle θ is at least 25.9°. (ii) Show that this will happen whatever the value of θ if the prism is entirely surrounded by air.

(b) (i) Find the orientation of the transmitted image of the arrow AB when the prism has been rotated about the incident beam axis (in air) by $\frac{1}{2}$ and $\frac{1}{4}$ of a revolution. (ii) What would you expect to see if the prism were rotated continuously and θ and the prism dimensions were such that the axial ray is ultimately collinear with its initial direction? [H a(i)]

H INTERFERENCE AND DIFFRACTION

H1 (*a*) A pair of narrow parallel slits is illuminated by monochromatic light of wavelength 500 nm to produce Young's fringes. It is found that, when one of the slits is covered by a thin film of transparent, non-dispersive material of refractive index 1.60, the zero-order bright fringe moves to the position previously occupied by a bright fringe of 15th order. (i) What is the thickness of the film? (ii) In which direction does the fringe pattern move? (iii) How could the zero-order fringe be identified?

(*b*) Take the slit separation as 1.0 mm and the distance of the observation screen as 2.0 m. With no film present the slits are illuminated by parallel white light (400–700 nm). Denoting wavelengths of 400, 500, 600 and 700 nm by B, G, Y and R respectively and white light by W, indicate approximately the colouration of the observation screen up to a distance of 4 mm from the centre of the observed pattern. [H *a*(i), (iii); H *b*]

H2 (*a*) A small sound transmitter T radiates uniformly in all directions and at four times the power of each of two other similar small transmitters, S_1 and S_2, placed 0.25 m on either side of T along a North–South line. The central transmitter is wired to be out of phase with the other two, and all three emit a 200 kHz signal. A small receiver R is placed 10 m due East of T and slowly moved eastwards. Where will the maximum and minimum signal responses occur? Take the velocity of sound in air as 350 m s^{-1}.

(*b*) At the position of a maximum response, (i) the two outer transmitters, and (ii) the central transmitter, are switched off for a while. By what factor does the power received fall in each case? [H *a, b*]

H3 A ship carrying an aerial at the top of its 25 m mast is transmitting on a wavelength in the range 2 to 4 m to a receiving station situated 150 m above sea-level at the top of a cliff. When the ship is 2 km from the foot of the cliff radio contact is lost. Find the radio wavelength used. Assume that the sea reflects radio waves perfectly. [H]

H4 A wedge-shaped air film is formed between two thin parallel-sided glass plates by means of a straight hair. The two plates are in contact along one edge of the film; the hair is parallel to this edge and 500 mm from it. The wedge is illuminated by light of wavelength 500 nm, incident normally from above, and the observed distance between the fifth and twenty-third bright fringe is 225 mm.

(i) What is the diameter of the hair?

(ii) Will the plates be light or dark at a distance of 250 mm from the edges in contact? [H (i), (ii)]

H5 A block of material has two parallel faces 10 mm apart and rests on one of them. A plane glass plate is placed over it, supported by two copper blocks which are just over 10 mm high, leaving a thin air film between the lower glass surface and the upper surface of the material. Interference fringes are observed through a microscope when sodium light of wavelength 589 nm is reflected normally from the glass plate. The cross-wires of a microscope are focussed on a dark fringe, and, when the temperature of the whole system is raised by 100 K, twenty bright fringes move past the cross-wires. What is the coefficient of expansivity of the material? The coefficient of linear expansivity of copper is $1.40 \times 10^{-5} K^{-1}$. [H]

H6 When the spectrum of light containing only a red and a violet component is examined with a diffraction grating (using normal incidence) of 300 lines mm^{-1}, it is found that a line at 24.46° contains both red and violet components.

(i) At what other angles, if any, would composite lines be found?

(ii) At what angles would pure red lines be found? [H]

H7 A long row of identical radio transmitters, spaced a distance d apart, is operating at a wavelength λ, but with signals from alternate aerials in antiphase (i.e. with a relative phase shift of π).

(i) Determine the direction(s) in which strong signals are transmitted.

(ii) Relate your answer to the situation in which alternate transmitters are switched off.

H8 (*a*) Estimate the greatest distance at which the two headlamps of a car could be distinguished from each other.

(*b*) The objective of an astronomical telescope for visual observation has a diameter of 1 m. What is the approximate magnification of the telescope that is best matched to this aperture size?

Note that the structure of the retina of the eye is such that the resolving power is limited to 2 minutes of arc whatever the pupil diameter. [H *a*, *b*]

H9 (*a*) Two transmitters 3 m apart emit equal amplitudes of 30 mm microwaves in a direction normal to the line joining the transmitters, and so produce an interference pattern 100 m away. Describe the interference pattern (i) if the two transmitters emit in phase, (ii) if the left-hand transmitter L is $\frac{1}{2}\pi$ ahead in phase, and (iii) if the left-hand transmitter emits at a frequency 1 Hz greater than the right-hand one R.

(*b*) What happens if the transmitters emit, in phase, waves which are plane-polarized, one parallel to the line joining the transmitters and one at right-angles to it? [H *a*(ii), (iii); H *b*]

H10* A house, situated 100 m to the South of a straight road running East–West, contains a television set which is receiving signals from a distant transmitter, also situated South of the road, operating at a frequency of 60 MHz. A bus, travelling along the road in a westerly direction, causes the received signal to fluctuate in intensity, the rate of fluctuation being 2 Hz when the bus is opposite the house and (temporarily) reducing to zero when it is 200 m further along the road. Find the speed of the bus and the direction of the transmitter. [H]

H11 Light of wavelength 590 nm falls at an angle of 45° on a thin soap film of refractive index 1.35. Dark bands are observed on the film 4.0 mm apart. Calculate the (small) angle α between the faces of the film. [H]

H12 A light source gives two lines of equal intensity. The wavelength of one line is 450.0 nm and that of the other slightly greater. The source is used to produce Newton's rings between a plane surface and a lens that can be supported at various distances above the surface. With the lens and surface in contact the rings are found to decrease in sharpness out to the 250th bright ring and then to increase in sharpness as the ring number is further increased. Calculate the distance(s) of lens–surface separation for which the central fringes disappear. [H]

H13 (*a*) Show that if the diffraction pattern of a single slit, of width a and illuminated by light of wavelength $\lambda(\ll a)$ is formed on a screen a (large) distance

$D(\gg a)$ from the slit, the first minimum of the (Fraunhofer) pattern occurs at a distance $\lambda D/a$ from the central maximum.

(*b*) A radar speed meter is situated 15 m from the side of a road, its beam making an angle of 15° with the road. If the transmitting aerial has a (horizontal) width of 0.20 m and the wavelength used is 30 mm, over what distance along the road can vehicles be detected? [H *a*, *b*]

I STRUCTURE AND PROPERTIES OF SOLIDS

I1 A fatty acid of relative molecular mass 282 and density 900 kg m^{-3} is introduced gradually onto a liquid surface of area 0.2 m^2 until the surface tension of the liquid suddenly starts to fall. Assuming that at this point a continuous layer of molecules lying flat on the surface has formed, and that a molecule is a chain about ten times as long as it is wide, estimate the mass of fatty acid added. [H]

I2 Electrons of energy 10 keV strike a tungsten target. The resulting X-rays are allowed to fall on a crystal which has a simple cubic structure with cube side equal to 5×10^{-10} m. It is found that, as the crystal is rotated, no X-rays are reflected unless the angle between the incident and reflected rays is greater than a certain value θ_0. Calculate θ_0. [H]

I3 The atoms of gold are close-packed, so that a representative unit consists of atoms at the corners of and in the middle of the faces of a cube. The planes on which the atoms are hexagonally arranged, having a normal along the body diagonal of this cube, are used to scatter X-rays through an angle of 37.0°. Given that the density of gold is 19.3×10^3 kg m^{-3}, and that the mass of one mole is 0.197 kg, what are the possible wavelengths of the radiation used? [H]

I4* A long-chain molecule is composed of identical atoms evenly spaced. The potential energy of interaction (between nearest neighbours only) is given by

$$V(x) = -\frac{A}{x^6} + \frac{B}{x^{12}},$$

where x is the distance between the two atoms. Calculate (i) the equilibrium spacing x_0 of the atoms, and (ii) the modulus of elasticity of the chain, both in terms of A and B. (iii) If the chain is gradually stretched at what strain will it break? [H (i), (ii), (iii)]

I5 In terms of the variation of interatomic potential energy with the distance between two atoms, explain (i) why crystalline solids normally expand with increasing temperature, and (ii) why the Young modulus for a crystalline solid decreases as the temperature increases.

I6 Sketch typical stress–strain (σ, ϵ) diagrams for the following materials (i) copper, (ii) glass, (iii) rubber, (iv) chewing gum, before chewing, and (v) chewing gum, after chewing.

I7 A composite material consists of parallel carbon fibres in a copper matrix and contains a fraction x, by volume, of fibres. Explain why for x less than a certain value x_0 the tensile fracture strength of the composite $\bar{\sigma}_t$ *decreases* as x increases, and calculate the value of x_0. The tensile fracture stress of copper σ_{mt} is 206 MN m^{-2} and its flow stress σ_{mf} is 41 MN m^{-2}. Carbon fibres have a tensile fracture stress σ_{ct} of 2000 MN m^{-2}. [H]

I8 A balloon ascends vertically, slowly unreeling a long copper wire. When 1 km of (unstretched) wire has been unreeled the wire breaks at the balloon. Estimate the amount by which the wire had stretched just before it broke. The density of copper ρ is 9×10^3 kg m^{-3}, and its Young modulus E is 1.2×10^{11} N m^{-2}. [H]

I9 A thin ring of radius R is made of a material which has density ρ and Young modulus E. If the ring is rotated about its centre, and in its own plane, with angular velocity ω, find the (small) increase in its radius. [H]

I10 Two similar rods, each of length L but of different materials, have Young moduli E_1 and E_2 and linear expansivities α_1 and α_2 respectively. They are fixed end-to-end and the combined rod is clamped rigidly at its ends. The temperature of the combined rod is now increased by ΔT. Find how far (x) the junction point shifts from its original position. [H]

I11* A hammer with a heavy metal head of mass m has a handle of length l, cross-sectional area A, but negligible mass, and is dropped handle first through a height h onto a rigid surface. During the impact both the head and the surface remain rigid, but the handle deforms elastically with Young modulus E. If $m = 2.0$ kg, $h = 1.0$ m, $l = 0.25$ m, $A = 1.2 \times 10^{-3}$ m^2 and $E = 1.5 \times 10^{10}$ N m^{-2}, find (i) an expression for the velocity v of the hammer head as a function of time during the impact, (ii) the duration τ of the impact, (iii) the maximum force F_m exerted at the point of impact, and (iv) the maximum compressive strain e_m in the handle. Note that the compression of the handle caused by the weight of the hammer head is about 10^{-6} m and can be ignored. [H]

I12 (*a*) A 20 kg weight is suspended from a length of copper wire 1 mm in radius. If the wire breaks suddenly, by how much does its temperature change?

(*b*) It has been suggested that the onset of plastic flow in a pure metal wire occurs when the stored elastic energy in the wire is equal to the latent heat of melting of the wire. On this basis estimate the elastic strain for which yielding would occur in copper, and compare your result with the practical value.

For copper: Young modulus $= 1.2 \times 10^{11}\,\mathrm{N\,m^{-2}}$, density $= 9.0 \times 10^{3}\,\mathrm{kg\,m^{-3}}$, specific heat capacity $= 0.42\,\mathrm{J\,kg^{-1}\,K^{-1}}$, specific latent heat of melting $=$ $1.1 \times 10^{5}\,\mathrm{J\,kg^{-1}}$. [H *a, b*]

I13 When a certain material is compressed the plot of fractional compression versus pressure is a straight line L from the origin to the point ($10^{7}\,\mathrm{N\,m^{-2}}$, 0.08). At this point the material yields and the compression increases indefinitely with no increase in pressure. If the pressure is subsequently reduced, the plot is a straight line parallel to L. A mass of 16 kg, with a light block of this material, of thickness 20 mm and area $0.2\,\mathrm{m^2}$, mounted in front of it as a buffer, collides with a similar block rigidly mounted.

(i) Find the maximum initial speed v_0 of the mass for which the collision is elastic.

(ii) Show that for speeds v_i greater than v_0 the rebound speed v_r is independent of v_i. [H (i)]

J PROPERTIES OF LIQUIDS

J1 A man in a boat on a pond of area A has with him a block of stone, more dense than water, and a block of timber, less dense than water. What happens to the level of the water in the pond if he throws out

(i) the block of stone,

(ii) the block of timber,

(iii) both of them tied together, so that they go to the bottom of the pond?

J2 A cartesian diver consists of a thin-walled cylinder, closed at one end, floating open end downwards in a liquid of density ρ. The small amount of air trapped in the cylinder is adjusted so that the diver floats just below the liquid surface when external pressure is P_0. The external pressure is then suddenly increased to $2P_0$. Show that in the subsequent motion of the diver, when it is at a depth x below the surface its speed v is given by

$$v^2 = 2gx - \frac{2P_0}{\rho} \ln\left(1 + \frac{\rho g x}{2P_0}\right)$$

Ignore any viscous effects and the density of air (compared with ρ), and assume that the air obeys Boyle's law. [H]

J3 A pond of density ρ is covered to a depth $\frac{1}{2}b$ by an oil of density $\rho'(<\rho)$. A long stick of square cross-section $2b \times 2b$ and density ρ' floats in the pond with one of its long faces horizontal.

(i) What fraction of it is immersed?

(ii)* By considering a situation in which the upper surface is inclined to the horizontal at a small angle θ and determining both the 'capsizing' and 'restoring' couples, show that the stick is stable provided $45\rho' > 37\rho$. [H (ii)]

J4 A homogeneous sphere of radius a, density ρ and negligible expansivity, floats in a liquid at 0 °C with its centre a height h above the level of the liquid which has cubic expansivity α. The surface tension σ of the liquid, which has zero angle of contact, varies with temperature t °C as $\sigma(t) = \sigma_0(1 - kt)$. Show that for a small rise in temperature of the liquid, the fraction of the sphere immersed does not change provided that

$$\frac{k-\alpha}{\alpha} = \frac{2a^4\rho g}{3\sigma_0(a^2-h^2)}.$$

[H]

J5 A rectangular framework has as two of its opposite sides rigid wires 20 mm long. The other two sides are light rubber threads, each 200 mm long, which obey Hooke's law and have force constants of 15 N m^{-1}. The wires are arranged in one plane so that the threads are initially straight but unstretched. When a soap film is formed in the framework it is found that the threads just touch at their centres. What is the value of the surface tension γ of the film? [H]

J6 A tube of internal diameter 8 mm is clamped in a vertical position. A second tube of 3 mm bore and 1.5 mm wall-thickness is suspended from the arm of a balance so that it hangs coaxially with the first tube, the lower ends of the two tubes being at the same level. A large vessel of water is then brought up so that the water's surface touches the lower ends of the tubes. A weight is then added to one of the balance scale-pans to restore the inner tube to its initial position.

(i) What is the difference in height between the water levels in the inner and outer spaces?

(ii) What weight was added?

(iii) Should your answer to (ii) be corrected for an Archimedean upthrust on the inner tube?

Take the surface tension of water as 72 mN m^{-1} and its angle of contact as 0°.

[H (i)]

J7 Explain with the aid of suitable diagrams why two floating objects are attracted to each other as a result of surface tension effects, whether they are floating on water or on mercury.

J8 Two soap bubbles of radii a and b are formed in an external pressure P, and then coalesce to form a bubble of radius c. Supposing that the air in the bubbles behaves as a perfect gas, and that its temperature does not change when coalescence occurs, show that

$$P = \frac{4\gamma(c^2 - b^2 - a^2)}{a^3 + b^3 - c^3},$$

where γ is the surface tension of the soap film. [H]

J9 The molecules of an organic liquid of density ρ are disc-shaped of radius r and thickness $0.1\,r$. X-ray diffraction studies suggest that in the liquid state the molecules are arranged in orderly fashion; close-packed edge-to-edge in a given plane with planes arranged so that the molecules in adjacent planes are packed face-to-face. The binding energy of two of these molecules face-to-face is ϵ, whereas edge-to-edge it is $\epsilon/15$.

(i) Find the specific latent heat l of evaporation of the liquid.

(ii) Determine whether molecules near the surface of the liquid will tend to be aligned with their faces parallel to or normal to the surface.

[H (i), (ii)]

J10 A uniform disc of mass M and radius a is suspended so that it can rotate freely about its own axis (with moment of inertia $\frac{1}{2}Ma^2$) with the plane of the disc a distance d from a large parallel plate. The space between the disc and the plate is filled with a liquid of viscosity η. Initially the disc is spinning with angular velocity ω_0. Find its angular velocity a time t later. Assume that the velocity gradient across the fluid is uniform. [H]

J11 (*a*) Show on dimensional grounds that the volume rate of flow $\mathrm{d}V/\mathrm{d}t$ of a liquid of viscosity η flowing through a horizontal cylindrical tube of radius r is related to the pressure gradient ϕ across the ends of the tube by a formula of the form

$$\frac{\mathrm{d}V}{\mathrm{d}t} = \frac{K\phi r^4}{\eta},$$

where K is a dimensionless constant (of value $\pi/8$).

(*b*) Two tall cylindrical vessels of radii a and b are joined near their bases by a horizontal narrow tube of length l and internal radius r ($a, b \gg r$). Initially they are filled to different depths with a liquid of density ρ and viscosity η. Assuming the subsequent flow to be slow, find the time required for the difference in levels, z, to fall to half of its original value. [H *b*]

J12 Poiseuille's formula for liquid flow (Question J11*a*) can be used for a compressible ideal gas if it is multiplied by a factor of $(P_1 + P_2)/2P_1$ for $\mathrm{d}V_1/\mathrm{d}t$, or by $(P_1+P_2)/2P_2$ for $\mathrm{d}V_2/\mathrm{d}t$, where the subscripts 1 and 2 indicate conditions at the two ends of the tube.

A soap bubble of radius $r_0 = 40$ mm and surface tension 30 mN m^{-1} is blown at the end of a glass tube of length 100 mm and internal diameter 2 mm. The viscosity of air is 1.85×10^{-5} kg m^{-1} s^{-1}. Find the time taken by the bubble to halve its radius. Note that the correction factor mentioned above is negligible in this case. [H]

J13* Oil of viscosity η and density ρ flows downhill in a flat shallow channel of width w and sloped at an angle θ. If the oil is everywhere of depth $d(\ll w)$ calculate the volume flow rate. Ignore viscous effects at the side walls. [H]

K PROPERTIES OF GASES

K1 One gram of helium at 20 °C is heated to 100 °C at constant pressure and then returned to 20 °C at constant volume. Calculate (i) the net heat absorbed by the helium in this process, and also (ii) the heat that would have been absorbed if an isothermal transition had been made between the same initial and final states. Assume helium to be an ideal gas. [H (i), (ii)]

K2 A thermally insulated container initially holds N_0 molecules of an ideal monatomic gas at an absolute temperature T_0. Molecules escape from the container through small holes in the walls, and it can be shown that in such a process at a temperature T the average kinetic energy of the escaping molecules is $2kT$. How many molecules remain in the container when their temperature has fallen to $\frac{1}{2}T_0$? [H]

K3 Modern central heating systems are sometimes permanently sealed, and instead of an expansion tank they have a closed bottle containing a quantity of air; the expanding water enters the bottom of the bottle, which is situated at the lowest point of the system, and compresses the air. The system is filled initially by isolating the bottle and then running cold water into the system, letting air escape at the top until the system is full. The system is then sealed at the top and the valve connecting the bottle to the system is opened. From the following information estimate the minimum size of the bottle.

Height of system = 15 m;

Volume of system = 0.20 m^3;

Operating range 5-100 °C;

Maximum safe pressure above atmospheric in the system = 3 atmospheres;

Cubic expansivity of water = 2×10^{-5} K^{-1}.

[H]

K4 A long uniform tube, open at one end, is surrounded by a gas at pressure p. The tube is heated so that its temperature varies uniformly from 1000 K at one end to 200 K at the other. The tube is then closed and cooled to a temperature of 100 K. Calculate the final pressure in the tube. [H]

K5 Two cylinders, A and B, each of volume V and negligible thermal capacity, are connected by a (closed) valve and in thermal contact with each other. Cylinder B is closed, but A is fitted with a piston which is initially fully withdrawn. Cylinder A contains a perfect monatomic gas at temperature T, whilst B is empty.

The valve is opened slightly and gas is driven as far as it will go into B by pushing home the piston at such a rate that the pressure p in A remains constant. Find the final temperature T_f of the system. [H]

K6 A balloon filled with 10^{-4} kg of hydrogen at a pressure of 1.05 atmospheres (1.05×10^5 N m^{-2}) is allowed to rise to the ceiling of a room. What area of its surface is in contact with the ceiling? Neglect the weight of the fabric of the balloon and take the mean relative molecular mass of air as 28.8. [H]

K7 On a certain day the temperature of the atmosphere is 15 °C at sea-level and decreases linearly with increasing height at a rate of 0.0065 °C m^{-1}. Treating air as a perfect gas of mean relative molecular mass 28.8 find the height x above sea-level at which the atmospheric pressure is half that at sea-level. [H]

K8 An electrically driven compressor delivers air at the rate of 10 litres s^{-1} and at 10 atmospheres pressure. If the overall efficiency of the motor and compressor is 70%, what is the power consumed from the mains? Assume the compression to be adiabatic and take γ to be 1.4 for air. [H]

K9 A simple air gun consists of a uniform tube of cross-sectional area A and length l, to one end of which is attached a closed chamber of volume V_0 containing air at pressure p_1, greater than the atmospheric pressure p_0. In the tube, immediately next to the chamber, a close-fitting pellet is placed, and the gun is discharged by making connection between the compressed air chamber and the tube. Ignoring air-leakage and friction, derive expressions for (i) the value of l which allows the greatest kinetic energy to be imparted to the pellet, and (ii) the magnitude of this energy (in terms of V_0, p_1, p_0 and γ). [H (i)]

K10* (*a*) A large, very light, balloon is partially filled with helium at ground level, where the atmospheric pressure is p_0 and the temperature $T(0)$. Assuming that the surface of the balloon never becomes taut and that the ascent is so rapid that conduction of heat into or out of the balloon is negligible, find the pressure of the atmosphere $p_A(x)$ at the maximum height x that the balloon will reach, the temperature at that height being $T(x)$.

(*b*) If the atmosphere is isothermal, show that this pressure is about $7\times 10^{-3}p_0$. [H *a*, *b*]

K11 A gas mixture consists of 3 moles of oxygen and 2 moles of helium. Show that for adiabatic changes it has an effective value for γ of 31/21, provided that oxygen and helium can be treated as ideal gases with values for γ of 7/5 and 5/3 respectively. [H]

K12 In a simple extension of the ideal gas model, due to Van der Waals, the equation of state for 1 mole, $pV=RT$, is modified to read

$$\left(p+\frac{a}{V^2}\right)(V-b)=RT, \qquad (\dagger)$$

where a and b are positive parameters which vary from gas to gas. The term involving a attempts to allow for the fact that molecules close to the wall of the gas container, where the pressure p is measured, experience a net attractive force towards the body of the gas as a result of a deficit of neighbouring gas molecules on the wall side of them. The quantity b is a measure of the actual volume of the gas molecules.

(i) Show that for this equation of state, the solutions $V(>b)$ of $\mathrm{d}p/\mathrm{d}V=0$ satisfy the equation

$$V-b=\left(\frac{RT}{2a}\right)^{1/2}V^{3/2}.$$

(ii) By sketching on the same graph the function $V\to V-b$ and functions of the form $V\to\lambda V^{3/2}$, deduce that in general there are either two solutions or none, the actual number depending upon whether λ is less than or greater than a particular value λ_c.

(iii) Show that, for $\lambda=\lambda_c$, there is only one solution $V=V_c=3b$, and that T then has the value $T_c=8a/27bR$.

(iv) Use the above results and the asymptotic forms of (†) for large V, and for $V\approx b$, to sketch typical isothermal curves for the gas. Take V as abscissa and p as ordinate. Relate the forms of the isothermals to the existence of a *critical temperature* T_c for a gas, an associated critical volume V_c, and a corresponding critical pressure p_c.

(v) Deduce that according to Van der Waals' model

$$\frac{p_c V_c}{T_c} = \frac{3R}{8}$$

for all gases. [H (ii), (iii), (iv)]

K13 A gas obeys Van der Waals' equation of state

$$\left(p + \frac{a}{V^2}\right)(V - b) = RT.$$

(i) Show that to first-order in small quantities, Boyle's law is obeyed at a temperature $T_B = a/bR$.

(ii) Assuming this temperature to be near 0 °C, estimate the order of magnitude of the (fractional) second-order variation in pV that would be observed in an apparatus working at temperature T_B if the maximum attainable pressure is 100 atmospheres. Take the linear dimensions of the gas molecules to be about 10^{-10} m and the molar volume V_M at s.t.p. to be 2.24×10^{-2} m^3 mol^{-1}.
[H (ii)]

L KINETIC THEORY AND STATISTICAL PHYSICS

L1 In high vacua, pressure is indicated by a mass spectrometer which measures the number of molecules per unit volume present in the vessel. Such an instrument is calibrated so that the pressure of air at room temperature (300 K) is directly given by a dial reading. What correction factor must be applied to read the pressure of helium gas at 20 K, assuming both helium and air behave as ideal gases? [H]

L2 In an industrial process 1000 ball-bearings, each of mass 1 g, are collected in a level square tray of sides $L=2$ m and have a total translational energy of 100 J. The balls suffer effectively-elastic collisions and move in random directions.

(i) What average force is exerted on one side of the tray?

(ii) If an opening of length $d=10$ mm were made in one side of the tray, estimate the time taken for the number of balls in the tray to fall to 500.

(iii) Why, with the conditions given (and for reasons not connected with statistical fluctuations), is (i) an exact calculation, but (ii) necessarily only an estimate? [H (i), (ii)]

L3 A sphere of radius r is lined with a metal which has a probability P of absorbing any gas molecule that hits it. If $P=\frac{1}{3}$, $r=5$ m, and the mean velocity of the gas molecules in the sphere is 400 m s^{-1}, find the time taken for the gas pressure p to fall to 1% of its initial value. [H]

L4 In a thermos flask the gas density is so low that the molecules hit the walls much more often than they hit each other. The flask walls may be treated as parallel and the temperature difference ΔT between the inside and outside is small.

(i) Obtain an approximate expression for the heat flux ϕ in terms of the gas pressure p, the average gas temperature T and the mass m of a molecule.

(ii) How does the flux depend on the wall separation d?

(iii) If N thermally isolated screens (which are free to 'float' in temperature) are placed between the walls (but not necessarily equally spaced) how, if at all, is the heat flux affected? Assume that on impact the molecules immediately come into thermal equilibrium with the walls or screens and that $\bar{c} = (8/3\pi)^{1/2} c_{\mathrm{r.m.s.}}$. [H (i), (iii)]

L5 A gas molecule of mass m and initial speed v_0 bounces perpendicularly back and forth between two walls of a container which are initially a distance x_0 apart. One of the walls is moved towards the other at a speed V very much less than v_0.

(i) Show that if the collisions are perfectly elastic then the product vx is constant, where v is the molecular speed when the wall separation is x.

(ii) Show that the average force experienced by either wall is $F = mv_0^2 x_0^2 x^{-3}$.

(iii) Show how the work done in compressing a thermally isolated gas is related to the increase in gas temperature. [H (i), (iii)]

L6 A piece of thin thermally insulating material $10^{-4}\,\mathrm{m}^2$ in area is held in helium, whose temperature is kept constant at 300 K. Light of intensity $10^2\,\mathrm{W\,m^{-2}}$ falls on one side of the material, which may be taken as a perfect absorber, and produces a small temperature rise on that side. Estimate the resultant force on the material.

(Assume that every molecule that impinges on the material leaves with a speed characteristic of the temperature of the surface, that all molecules arrive at and leave the surface normally, that radiation losses can be ignored and that there is a unique molecular speed corresponding to any particular temperature. Ignore the pressure due to the radiation.) [H]

L7 (*a*) A 'crystal' of four black atoms and a 'crystal' of four white atoms are placed in contact and allowed to reach equilibrium. If only one atom is allowed to occupy any one lattice site, what is the probability of finding three black atoms in either of the crystals?

(*b*) In a hydrogen molecule the pair of protons may be in any of four states (distinguished by proton spin orientations). In pure *para*-hydrogen, which is the equilibrium form at very low temperatures, all protons are in the same state, but in normal hydrogen at room temperature they are distributed between the four states at random. If one mole of pure *para*-hydrogen is brought to room temperature (300 K) and is then left to convert, slowly but irreversibly, into normal hydrogen, (i) by how much does its entropy change during the conversion process, and (ii) what can be said about the heat which is emitted or absorbed? [H *a*]

L8 A 20 Ω resistor has a mass of 5 g and is made of a material whose specific heat capacity is 800 J kg^{-1} K^{-1}. An electric current of 5 A flows for 2 s through the resistor under each of the following conditions.

(i) The resistor is kept at a temperature of 27 °C by a stream of running water.

(ii) The resistor is thermally insulated and initially at a temperature of 27 °C.

What are the changes in entropy of the resistor and of the water in case (i), and of the resistor in case (ii)? [H (ii)]

L9 If W is the number of ways in which a system can be arranged, estimate the ratio W_2/W_1 of the values of W after and before the following processes.

(i) The melting of 5 moles of ice.

(ii) The attainment of equilibrium when two 1 mole copper blocks, initially at temperatures of 0 °C and 100 °C, are brought into thermal contact.

The molar latent heat of melting of ice is 6×10^3 J mol^{-1} and the molar heat capacity of copper is 25 J mol^{-1} K^{-1}. [H (i), (ii)]

L10 Find the change in entropy which occurs when 2 moles of gas B and 1 mole of gas C, both at s.t.p. and in adjacent volumes, are allowed to mix by removing the partition between them. [H]

L11 (*a*) The volume of 1 mole of a perfect gas is doubled by two different processes, during each of which it is in thermal contact with a thermal reservoir at constant temperature T. In each case calculate the change in entropy of the gas and relate it to the heat that flows between the reservoir and the gas. (i) The gas is initially contained in a cylinder of volume V by a frictionless, tight-fitting piston, and is allowed to expand slowly, pushing back the piston. (ii) The gas is in a container of volume V when one wall of the container is suddenly removed and the gas expands into a second container of equal volume which was previously empty.

(*b*) (i) Show that if the second process is approximated by an adiabatic expansion followed by the return of the gas temperature to T, the heat that flows is

$$c_{v,m}T(1 - 2^{-R/c_{v,m}}).$$

(ii) Verify by numerical experiments which take $c_{v,m}$ as $(n+\frac{1}{2})R$ that this quantity approaches but does not exceed $RT \ln 2$ as $n \to \infty$. [H *b*(i)]

L12 The specific latent heat needed to evaporate water is about 2.3 MJ kg^{-1}. Estimate the temperature T at which an airing cupboard should be maintained in order to dry clothes ten times faster than it would at 27 °C. Indicate any simplifying assumptions on which the estimate depends. [H]

L13* Two identical bodies, each of constant heat capacity C, are at temperatures T_1 and T_2. What is the maximum amount of work which can be extracted by allowing them to reach equilibrium with each other? [H]

M HEAT TRANSFER

M1 A thermistor of base area A is mounted on a heat sink at temperature T_0 and separated from it by an insulating washer of the same area, thickness d and thermal conductivity λ. The electrical resistance of the thermistor is given by $R = a/T$ where a is a constant and T its temperature. If conduction through the washer is the only way of removing heat from the thermistor, and a voltage V is applied across the thermistor, (i) find an expression for T. (ii) What happens if $V^2 d > \lambda A a$?

M2 The surface of a large lake of water at 0 °C is maintained at −2 °C. Assuming that only thermal conduction is important, and using relevant data chosen from that given below, estimate how long it would take for a layer of ice 10 cm thick to form on the lake's surface.

Thermal conductivity of water = $0.56\ \mathrm{W\,m^{-1}\,K^{-1}}$.

Thermal conductivity of ice = $2.3\ \mathrm{W\,m^{-1}\,K^{-1}}$.

Specific latent heat of fusion of ice = $3.3 \times 10^5\ \mathrm{J\,kg^{-1}}$.

Density of water = $1000\ \mathrm{kg\,m^{-3}}$.

Density of ice = $917\ \mathrm{kg\,m^{-3}}$. [H]

M3 If it takes two days to defrost a frozen ten-pound turkey, estimate how long it would take to defrost a two-ton Siberian Mammoth. [H]

M4 A Y-structure made of copper has three identical limbs of length l, diameter d and thermal conductivity λ. Two of them are connected to a gold block of thermal capacity C at their free ends, whilst the free end of the third limb is maintained at a temperature T_0. The whole apparatus is mounted in vacuo and radiation losses and the thermal capacity of the copper can be neglected.

(i) What is the temperature T of the centre point of the Y-structure when the temperature of the gold block is T_1?

(ii) How long does it take for the gold block to reach a temperature $\frac{1}{2}(T_2 + T_0)$ starting from a temperature T_2? [H; H (ii)]

M5* Due to fission of the atoms, heat is generated uniformly throughout the volume of a sphere of uranium of radius $a = 100$ mm at a rate of $H = 5.5 \times 10^3\ \mathrm{W\,m^{-3}}$. If the thermal conductivity λ of uranium is $46\ \mathrm{W\,m^{-1}\,K^{-1}}$, what is the temperature difference between the centre and the outside of the sphere in the steady state? [H]

M6 (*a*) Show that the rate of flow of heat per unit length ϕ, through the walls of a thick-walled tube of internal and external radii r_1 and r_2 respectively and made of material of thermal conductivity λ, is given by $\phi = 2\pi\lambda/\ln(r_2/r_1)$ when the temperature difference between the inner and outer surfaces of the tube is unity.

(*b*) A fire-tube boiler consists of a number of such tubes conveying gaseous combustion products from furnace to flue. Heat flows radially by conduction, from the hot gases to water at sensibly constant temperature T_0 surrounding the tubes. The gas flows in the tubes without appreciable pressure drop at a mass flow rate f_m and with constant specific heat capacity c. Show (i) that the temperature $T(x)$ of the gas at a distance x from the entry end of the tube (where the temperature is T_1) satisfies

$$\frac{f_m c}{\phi}\frac{\mathrm{d}T}{\mathrm{d}x} + T = T_0,$$

and (ii) that, if the length of a tube is l, the overall heat transfer rate H from gas to water for that tube is given by

$$H = f_m c(T_1 - T_0)\,[1 - \exp(-\phi l/f_m c)].$$

[H *a*; H *b*(i), (ii)]

M7 A body of heat capacity C when warmed by a heater of power P can attain a maximum temperature T_1 if its surroundings are at a temperature T_0. Assuming that heat losses from it obey Newton's law of cooling (i.e. the rate of heat loss is proportional to the excess temperature over the surroundings), find the time taken for the temperature to fall to $\frac{1}{2}(T_1 + T_0)$ if the heater is switched off whilst the body is at temperature T_1. [H]

M8 The Sun's diameter subtends an angle of $0.35°$ when seen from Mars. Taking the Sun as a black body at a surface temperature of 6000 K and treating Mars as an absorbing black disc of area πr^2 upon which solar radiation falls normally, and as a sphere of area $4\pi r^2$ radiating into space at 0 K, estimate the surface temperature T of Mars. [H]

M9 A convex lens of diameter 100 mm and focal length 500 mm is used to produce a focussed image of the Sun on a thin matt black disc the same size as the image. What is the highest temperature to which the disc can be raised if the Sun has a black-body temperature of 6000 K? [H]

M10 A very large, flat, black sheet is maintained at 100 °C just above a similar sheet which is kept at 10 °C by a stream of water, the space between the sheets being evacuated. Three further sheets of black metal, thermally insulated from the first two and from each other, are now inserted between them and allowed to come into equilibrium.

(i) What temperatures will they take up?

(ii) What will be their effect on the rate at which heat is carried away by the water? [H (i)]

M11 A 60 W fluorescent lamp has a surface area of 0.1 m^2 and emits 3% of its input as visible radiation. When it is running the measured temperature of the fluorescent powder is 350 K; for this temperature about 0.1% of a black-body emission spectrum is in the visible range. Demonstrate and explain the apparent inconsistency of these measurements.

M12 An electric furnace containing platinum at its melting point is used as a standard of radiation intensity. The furnace wall has an area of 0.12 m^2 in which there is an observation hole of area $10^{-3} m^2$, the remainder being covered with refractory brick 20 mm thick. The outside of the brick is at a temperature of 200 °C. What electric power is required to maintain the temperature of the furnace? The m.p. of platinum is 1773 °C and the thermal conductivity of refractory brick 0.16 $W m^{-1} K^{-1}$. [H]

M13 A uniform metal bar of length l and thermal conductivity λ has one of its end faces maintained at temperature T_1 and its sides lagged so that no significant amount of heat is lost through them. The other end face behaves as a black body and loses heat by radiation to its surroundings which are at a constant temperature T_0. Show that in the steady state the temperature T of this face is given approximately by

$$T = \frac{\lambda T_1 + 4\sigma T_0^4 l}{\lambda + 4\sigma T_0^3 l},$$

where σ is the Stefan–Boltzmann constant and $T - T_0$ is assumed to be $\ll T_0$. [H]

N ELECTROSTATICS

The capacitances of simple capacitors are given by the following formulae: parallel plate, $\epsilon_0 A/d$; concentric spheres, $4\pi\epsilon_0 ab/(b-a)$; isolated sphere $4\pi\epsilon_0 a$.

N1 Two parallel conducting plates each of area A and with separation d are connected to a source of constant voltage V. The plates are allowed to approach each other gradually until their separation is $\frac{1}{3}d$. The source is then disconnected and the separation of the plates is slowly restored to the value d. What difference, if any, is there between the initial and final electrostatic energies stored in the capacitor? [H]

N2 The plates of a parallel-plate capacitor each of area A and with separation x are charged to a potential difference V by a battery. Calculate the change in the stored energy if the plates are pulled apart by a small amount dx (i) with the battery connected, (ii) with the battery disconnected. (iii) Explain any difference between the results for (i) and (ii). (iv) What force is necessary to pull the plates apart in the two cases?

N3

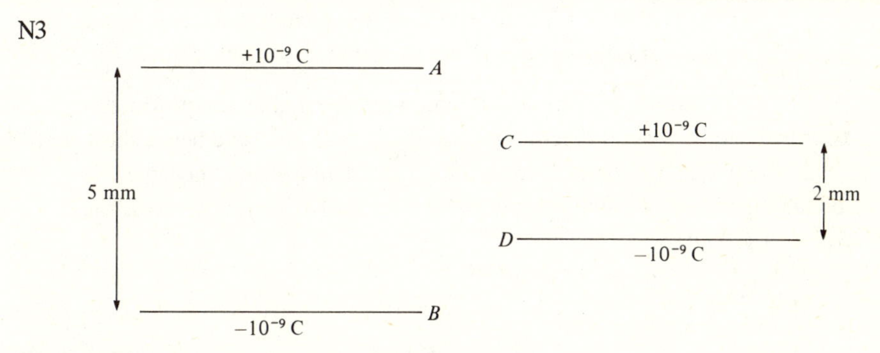

The diagram shows two parallel-plate capacitors in cross-section, each carrying a charge of 10^{-9} C with signs as shown. The left-hand one has a capacitance of 2×10^{-11} F. The four plates A, B, C and D are identical, so the two capacitors differ only in the separations of their plates. What are the potential differences

V_{AB} and V_{CD} (i) initially, (ii) if B is connected to D, (iii) if B is connected to D and A to C, (iv) if A is connected to D and B to C, and (v)* if the right-hand capacitor is slid between the plates of the left-hand capacitor, without touching them? Assume that the plates are very thin. [H (v)]

N4 An air-filled capacitor consists of three parallel metal plates each of area $10^{-2}\,\text{m}^2$. The distances between the inner plate and the two outer ones are 1 mm and 2 mm. Initially the outer plates are both earthed and the central one is charged to a potential of 3000 V. All three plates are now insulated and the central one removed. Calculate (i) the charges left on the outer plates, and (ii) the final potential difference between them. [H (ii)]

N5 Two isolated conducting spheres, each of radius 0.1 m, are charged to 200 V and 400 V and then connected by a fine wire. Calculate the heat generated in the wire. [H]

N6 A point charge q is placed at a point P a distance b from the centre C of an earthed conducting sphere of radius $a(<b)$. It can be shown that the potential V *everywhere* in the space outside and on the sphere is the same as would arise from just two charges (with the sphere removed); the original charge q at P and a second charge of magnitude $-qa/b$ placed between C and P, and on the line joining them, a distance a^2/b from C.

(i) Verify that these two charges do give $V=0$ at the two points at which the line CP meets the sphere's surface, and also on the circumference of the sphere which lies in a plane to which CP is normal.

(ii) Verify that $V=0$ everywhere on the sphere's surface. (Consider a point X on the surface the radius to which makes an angle θ with CP.)

(iii) Consider the physical situation represented by the previously described pair of charges together with a charge qa/b at C, and hence show that a charge Q placed a distance R from the centre of an isolated uncharged conducting sphere of radius a experiences a force of attraction towards the sphere of magnitude

$$\frac{Q^2}{4\pi\epsilon_0}\left(\frac{a^3(2R^2-a^2)}{R^3(R^2-a^2)^2}\right).$$

[H (ii), (iii)]

N7 A conducting balloon of radius a is charged to a potential V. It is connected to earth through a resistance R, and at the same moment a valve on the balloon is opened.

(i) Show that if the balloon's radius r decreases at a certain constant rate the potential of the balloon remains constant.

(ii) How much heat is dissipated in the resistance under these circumstances? [H (i)]

N8 A battery consists of N identical cells (each of e.m.f. $\mathcal{E}$) connected in series. A capacitor is charged via a resistor (*a*) by connecting it across the terminals of the entire battery, and (*b*) by connecting it first across a single cell, then across two cells, and so on, until it is again fully charged. Show that the energy of the battery that is wasted (i.e. which does not go into the stored energy of the capacitor) is N times smaller in case (*b*) than in case (*a*). [H]

N9 The vertical electric field below a thundercloud is about $10^4\,\mathrm{V\,m^{-1}}$.

(i) What is the smallest charge a raindrop of radius 1 mm would have to carry in order for it not to fall under the action of gravity?

(ii) If the breakdown strength of air is approximately $3\,\mathrm{MV\,m^{-1}}$, determine whether it is likely that the drop could carry such a charge.

(iii) Suppose that such a drop could carry a sufficiently large charge that it would not fall under gravity. If it were to split into two smaller equal drops the surface energy would be increased by about 2.5×10^{-7} J because of the increased total surface area of water. Show that, despite this, the drop would tend to split in this way rather than remain intact. [H, (iii)]

N10 It has been suggested that a parallel-plate capacitor might be constructed in which the separation d between the plates varies with the charge Q stored on the plates according to the formula $d = d_0 - \alpha Q$ where d_0 and α are constants. Such a capacitor carries a total charge $Q = Q_0 + Q_1 \cos \omega t$, made up of a fixed charge and an oscillating charge supplied by an external circuit. Show that the voltage across the capacitor will have components of more than one frequency, and that the component of angular frequency ω can be made zero by a suitable choice of Q_0. [H]

N11 A rectangular box of volume v contains $2N$ thin conducting plates, each of area A, arranged parallel to one of the faces of the box. Alternate plates are electrically connected together and the spaces between the plates are air-filled. If the breakdown electric field strength for air is E_0, show that the maximum electrostatic energy that can be stored in the box is $\frac{1}{2}\epsilon_0 v E_0^2$, independent of the value of N used. [H]

N12* A spherical capacitor consists of an outer sphere of fixed radius b and a concentric inner sphere whose radius a can be chosen. The space between the spheres is filled with air which has breakdown electric field strength E_0. What are the greatest achievable values for (i) the potential difference between the spheres, and (ii) the electrostatic energy stored in the capacitor? [H; H (i), (ii)]

N13 (*a*) (i) Find an expression for the electrostatic energy (in vacuo) of a spherical charged conductor of radius a and uniform surface charge density σ. (ii) By finding the change in energy for a small change in radius, the total surface charge being held constant, show that the outward pressure, p, on the sphere is $\sigma^2/2\epsilon_0$.

(*b*) A thin-walled metal sphere of radius $a = 0.5$ m is charged. If the thickness of the metal is 1 μm, what must be the tensile strength of the material (in $\mathrm{N\,m^{-2}}$) if the sphere is not to rupture before discharging by ionizing the surrounding air? Air ionizes in an electric field intensity $E_0 = 3\ \mathrm{MV\,m^{-1}}$ and the excess pressure is related to T, the tensile force acting across a line of unit length in the surface, by $p = 2T/a$.

(*c*)* If the cost of making a spherical shell is proportional to the volume of material in the shell, is it better to use a number of small spheres or one big sphere (all of the same thickness) to store a given amount of electrical energy?

[H *a*(ii), *b*, *c*]

O DIRECT CURRENTS

O1 A rectangular electrolytic tank has glass sidewalls and silver endwalls. It is partitioned by a tightly fitting central copper plate parallel to the endwalls, and one-half of the tank contains silver nitrate solution ($AgNO_3$), whilst the other contains a solution of copper sulphate ($CuSO_4$). The two endwalls are connected across a constant d.c. potential of 100 V. The weight of the central plate is then found to increase at the rate of 8 mg s^{-1}. Calculate the power taken from the d.c. supply. The relative atomic masses of silver and copper are 108 and 63.5 respectively. [H]

O2 (*a*) Estimate the average drift speed v_d of the electrons in a copper wire of radius $r = 2$ mm carrying a current $I = 25$ A. Assume that each copper atom contributes one electron to the drift current, that the relative atomic mass of copper is 63.5 and that its density is 8.9×10^3 kg m^{-3}.

(*b*) Take as a model of ohmic loss that electrons undergo collisions (with atoms of the lattice) in which they lose all their kinetic energy, which then appears as heat. The electrons are then uniformly accelerated again from rest. By considering the power loss in a wire of length l and cross-sectional area A, show that the resistivity ρ of copper is given, on this model, by $2m_e/\tau ne^2$, where $\tau = 4 \times 10^{-14}$ s is the time between collisions and n the number of electrons per unit volume. Evaluate ρ. [H *b*]

O3 (i) A two-core underground cable of constant resistance per unit length ρ and 7 km long joins points A and B. A fault on the cable at some point along its length has the effect of connecting the two cores of the cable at that point with a resistance r. The resistances measured at A and B when the opposite ends of the cable are open-circuited are 64 Ω and 70 Ω respectively. When 16 V is applied at A the voltage (measured with a high impedance meter) at B is 15 V. Deduce the value of the resistance r and the position x along the cable at which the fault occurs.

(ii) What fraction of the power supplied at A would be absorbed in a load of 50 Ω connected at B? [H (i), (ii)]

O4 (i) A 1.3 V battery with an internal resistance of 1.5 Ω is connected in parallel with a 2 V battery having an internal resistance of 2.5 Ω. What power will be dissipated in a 3 Ω resistor connected across this combination?

(ii) Two d.c. generators have separate current–voltage characteristics as follows:

I/A	0	50	100	150	200
V_1/V	240	238	234	226	210
V_2/V	240	236	226	210	180

Determine the power dissipated in a 1 Ω resistor when both generators are connected to it in parallel. [H (ii)]

O5 What is the largest current which can be carried by a copper fuse wire of radius 1 mm? Ignore heat conduction and take the resistivity of copper to be 1.8×10^{-8} Ω m. Its melting point is 1080 °C. [H]

O6 Two physics students, A and B, living in neighbouring college rooms decided to economize by connecting their ceiling lights in series (to the d.c. supply). They agreed that each would install a 100 W bulb in his own room and that they would pay equal shares of the electricity bill. However both decided to try to get better lighting at the other's expense; A installed a 200 W bulb and B installed a 50 W bulb. Which student subsequently failed his examinations?

O7 A moving-coil galvanometer has a resistance of 1200 Ω and gives a full-scale deflection when there is a potential difference of 25 mV across the coil. How would you connect it, with external resistors, as (i) an ammeter measuring 10 amps full-scale, and (ii) a d.c. voltmeter measuring 100 volts full-scale? (iii) If both modifications are (accidentally) made at the same time, and the meter is connected across the output of a d.c. generator, what is the minimum (theoretical) power needed to cause a full-scale deflection?

O8 (*a*) Find under what conditions (i) maximum power, and (ii) maximum energy, can be delivered to an electric motor driven by a battery of finite storage capacity, of e.m.f. E and of fixed internal resistance R. The resistance r of the motor windings is the quantity which can be varied.

(*b*) A lead–acid battery consists of six cells in series each of e.m.f. 2.0 V and internal resistance 0.01 Ω. The motor windings are chosen so as to produce maximum power in the motor. Now one of the cells in the battery starts to develop a higher internal resistance than 0.01 Ω. What is the minimum value of the internal resistance R' of this cell such that for maximum power in the motor it should be short-circuited? [H a(ii); H b]

O9 Two (non-linear) devices A and B are connected in parallel and this unit is connected in series with a third (non-linear) device C. The resistance (defined as voltage/current) of A is aI_A, where a is a positive constant and I_A is the current through A; the resistances of B and C are similarly defined as bI_B and cI_C respectively. Show that, if a voltage source V is connected across the system described, the power drawn from the source is

$$\frac{(a^{1/2}+b^{1/2})V^{3/2}}{[ab+ca+cb+2c(ab)^{1/2}]^{1/2}}.$$

[H]

O10 A Wheatstone bridge consists of four conductors AB, BD, AC, and CD. A 2 V cell is connected between A and D, and a galvanometer of resistance $G=20\,\Omega$ between B and C. Initially all the arms of the bridge have resistances $R=10\,\Omega$. If the resistances of arms AB and CD are both increased by $r=1.0\times10^{-4}\,\Omega$, how much current will flow through the galvanometer? [H]

O11

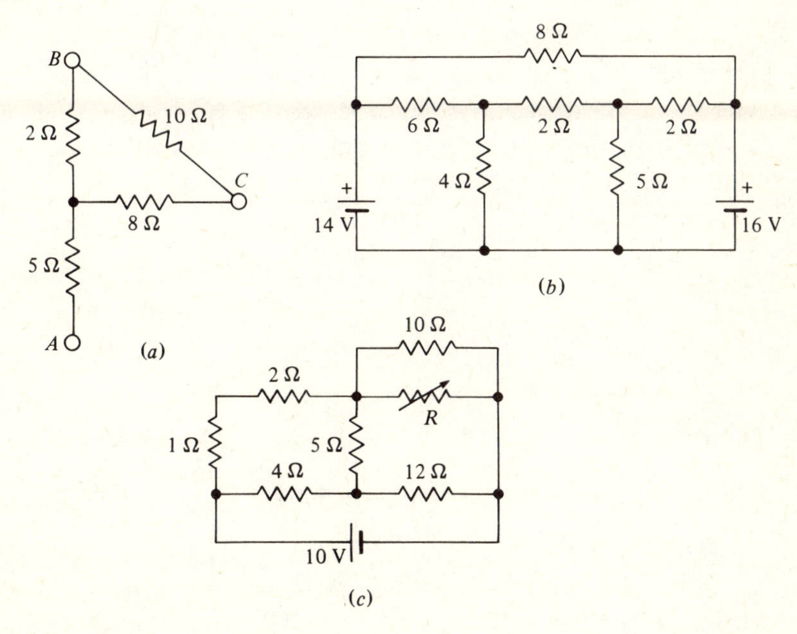

(*a*) For the network shown in figure (*a*) determine the resistance between the following pairs of terminals, (i) *A* and *B*, (ii) *B* and *C*, (iii) *B* and *C* when terminals *A* and *C* are shorted.

(*b*) Find the current in the 4 Ω resistor in the network shown in figure (*b*).

(*c*) For the circuit shown in figure (*c*), at what value should the variable resistor *R* be set in order to minimize the heat generated in the 5 Ω resistor?

[H *b, c*]

O12

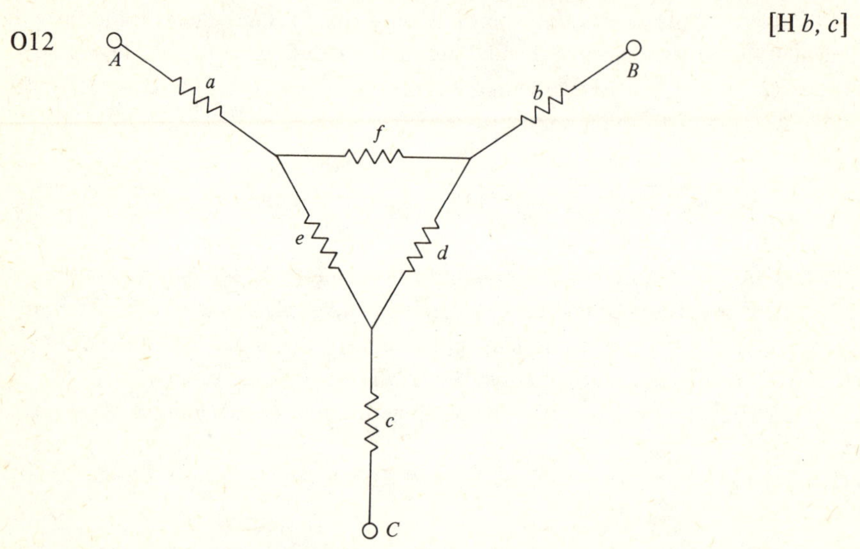

Six resistors with resistances of 1 Ω, 2 Ω, ..., 6 Ω are connected in the configuration shown. The resistances measured between the three pairs of terminals are: *AB*, $7\frac{3}{13}$ Ω; *AC*, $6\frac{9}{13}$ Ω; *BC*, $10\frac{1}{13}$ Ω. Determine which resistor is which. [H]

O13

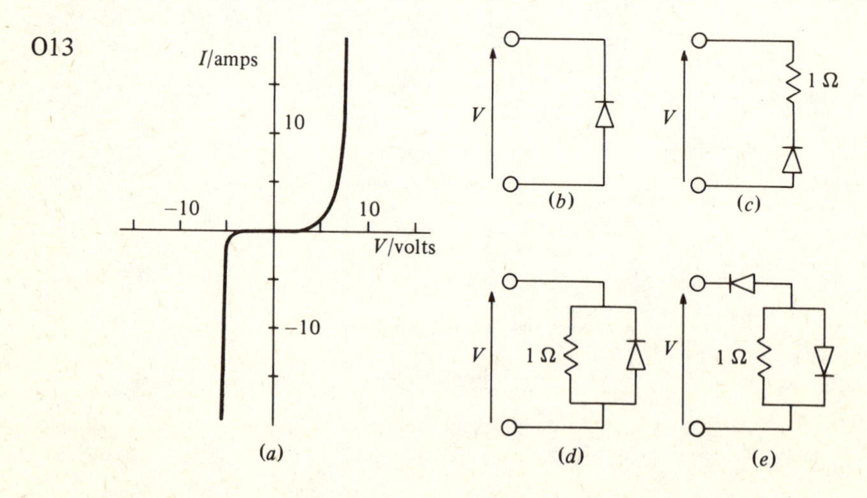

The rectifying diode shown in figure (*b*) has the current–voltage characteristic sketched in figure (*a*). Sketch the current–voltage characteristics for the diode–resistor combinations shown in figures (*c*), (*d*) and (*e*). Accurate drawings are not required.

P NON-STEADY CURRENTS

P1 Determine the maximum displacement a of an electron in a copper wire of diameter 1 mm which carries a 50 Hz alternating current of r.m.s. magnitude 10 A. The density of copper is $8.9 \times 10^3 \text{ kg m}^{-3}$, and its relative atomic mass 63.5. Assume that each copper atom contributes one electron to the current. [H]

P2 A digital a.c. voltmeter detects average absolute voltage, but, on the assumption of a sinusoidal voltage source, displays r.m.s. voltage. What reading will it display if it is connected to a voltage source which changes abruptly, at equal intervals of time, from +10 V to −5 V and vice versa? [H]

P3 (*a*)

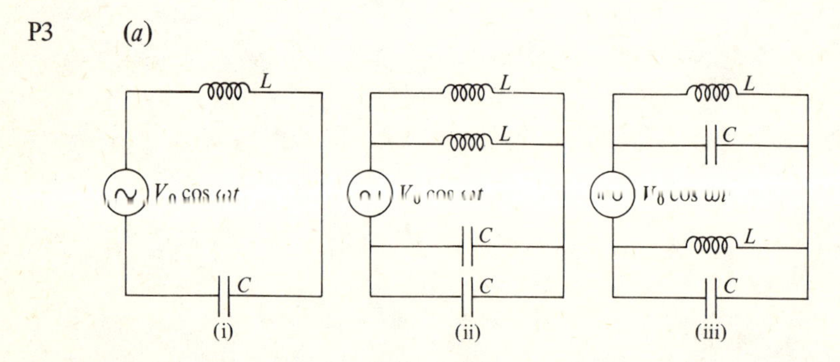

For the circuit of figure (i) assume that the current I is given by $I = I_0 \sin \omega t$, where I_0 may depend upon ω. Hence show that there is (current) resonance when $\omega = \omega_0$, where $\omega_0 = (LC)^{-1/2}$. For the circuits of figures (ii) and (iii), determine and sketch qualitatively the variation with ω of the magnitude of the current drawn from the source.

(*b*) Show by means of a series of circuit diagrams how, if you were supplied with only the components in figure (ii), you would construct five new circuits (different from those given in the figures), each of which would show (current) resonance at a different frequency. Label each diagram with the corresponding resonant frequency, expressed in terms of ω_0. [H *a*(ii), (iii)]

P4 A 75 W non-inductive electric light bulb is designed to run from an a.c. supply of 120 V (r.m.s.) and 50 Hz. If the only supply available is 240 V (r.m.s.) and 50 Hz, show that the bulb can be run at the correct power by placing either (i) a resistance R or (ii) an inductance L, in series with it. Find the values of R and L, and the power drawn from the supply in each case.

P5 A varying potential difference V_1 is applied across a resistance R and a capacitor C in series. The p.d. V_2 across the capacitor is recorded by a very high impedance voltmeter. Show that the variation of V_1 with time t can be deduced from a (t, V_2) plot as follows.

(i) Let P be the point on the curve corresponding to time t.

(ii) At time $t + RC$ draw an ordinate to meet the tangent to the curve at P in the point Q.

(iii) Draw QN parallel to the t-axis to meet the ordinate through P in the point N.

(iv) Then $V_1(t)$ is given by the locus of N. [H]

P6 A diode, which can be assumed to have negligible resistance when conducting in the forward direction, is used in the circuit shown to rectify a 20 V peak-to-peak a.c. supply of frequency $f = 50$ Hz.

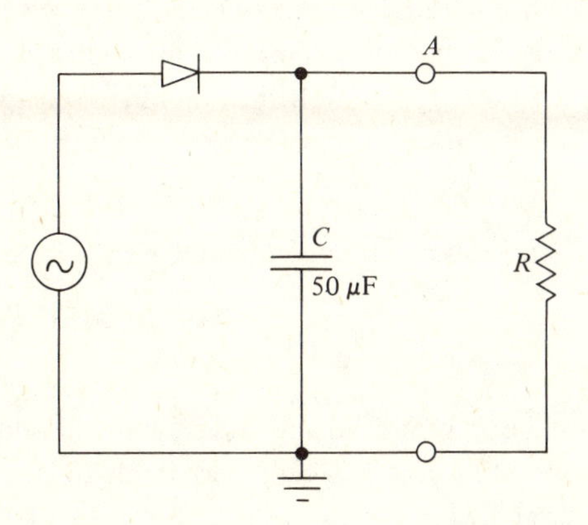

Sketch the waveforms which would be observed at the terminal A for values of the load resistance R of (i) 10 kΩ, and (ii) 100 Ω. (iii) Estimate the peak-to-peak ripple of the output voltage for case (i). (iv)* Show that in case (i) the diode is conducting for approximately $4\frac{1}{2}$% of the time. [H (iii), (iv)]

P7

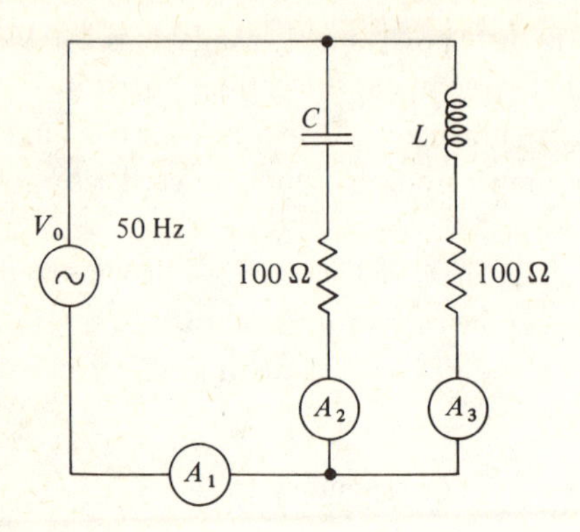

In the circuit shown the readings of the three a.c. ammeters A_1, A_2 and A_3 are all equal. What are the values of L and C? [H]

P8 The current I through a certain non-linear circuit element is given by the relationship $5I = V^2$, where I is measured in micro-amps and V, the applied voltage, in volts. The element is connected across a 4 μF capacitor which is initially charged to a p.d. of 100 V. How long will it take for this p.d. to fall to 50 V? [H]

P9 A resistance R and a capacitance C are connected in series with a low-impedance source of e.m.f. V_0. Across the capacitor is connected a device which has the property that, if initially non-conducting, it remains non-conducting until the voltage across it rises to the value $V_1 (< V_0)$. It then rapidly discharges the capacitor until the voltage across it drops to a small value, effectively zero; whereupon it returns to the non-conducting state.

(i) Show that the circuit generates a periodic potential V of period

$$T = RC \ln \left(\frac{V_0}{V_0 - V_1} \right).$$

(ii) Sketch the voltage waveform produced and choose suitable values of the parameters of the circuit so that the voltage would be suitable for an oscilloscope time-base, linear to 1% and with $T = 0.01$ s. [H (i), (ii)]

P10 Figure (i) shows a circuit which incorporates a (differential) amplifier whose output V_3 is either +10 V or −10 V, depending upon which of its two inputs is at a higher potential (as shown in figure (ii)). Both inputs take negligible current.

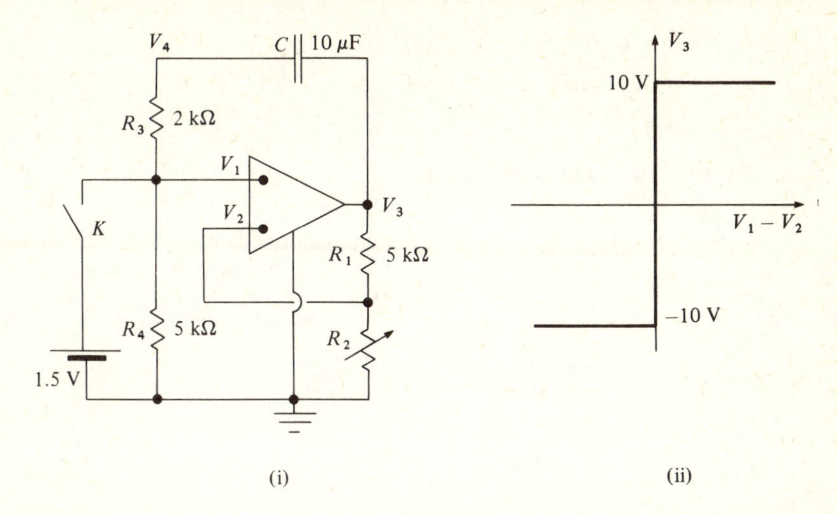

(i) (ii)

(*a*) Initially the capacitor is uncharged and R_2 is set to 0 Ω. The key K is closed momentarily and then left open. Determine the subsequent time variation of the current taken from the output of the amplifier.

(*b*) R_2 is now set to 1 kΩ. Without a full mathematical analysis show that the system becomes an oscillator of frequency about 5 Hz. Sketch the qualitative time-variation of V_1. [H *a*, *b*]

P11 (*a*)

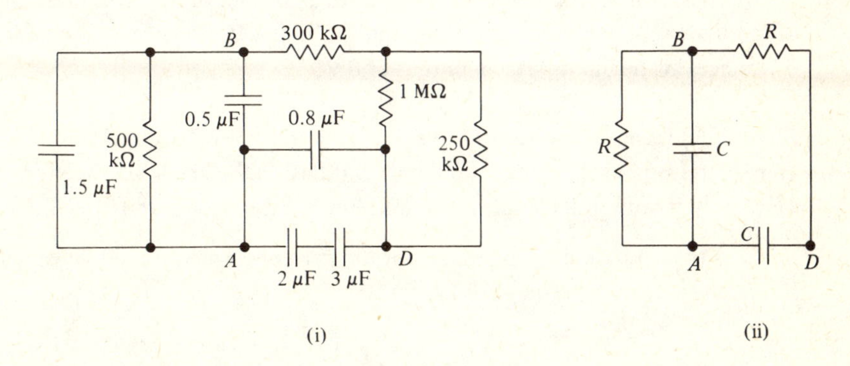

(i) (ii)

Show that the circuit in figure (i) is equivalent to the simpler one shown in figure (ii) so far as the voltages at A, B and D are concerned. Find the appropriate values of R and C.

(*b*) If the potential drop from A to B is denoted by V_1, and that from B to D by V_2, obtain two equations connecting V_1 and V_2 and deduce that

V_1 satisfies the equation

$$\tau^2 \frac{d^2 V_1}{dt^2} + 3\tau \frac{dV_1}{dt} + V_1 = 0,$$

where t denotes time. Find the value of τ. [H *a, b*]

P12 A transistor unit has two resistive inputs and one capacitive one, as shown in the diagram.

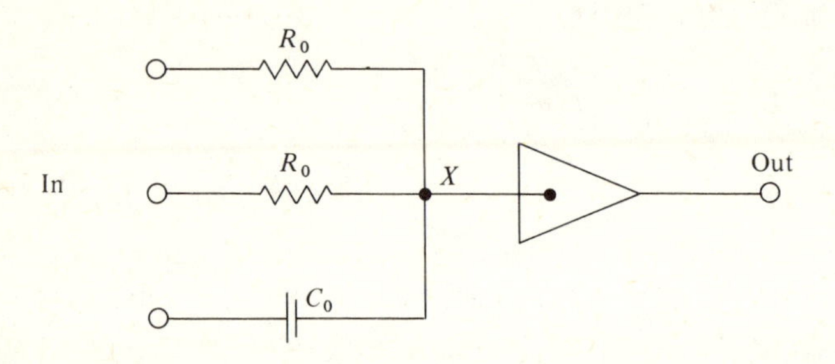

It has the characteristic that its output is 6 V if the voltage at point X is <3 V, but is 0 V if that voltage is >3 V. Using the minimum number of these units, construct the following systems:

(i) A two-input system which has an output of 0 V if both of the inputs are at 0 V, but has an output of 6 V otherwise.

(ii) A three-input system which gives an output of 6 V whenever inputs A and B are simultaneously at 6 V but input C is at 0 V. All other combinations of voltages at the inputs give an output of 0 V. [H]

P13 (i) Using the transistor units described in **P**12 and two 6 V lamps construct a system which has two inputs and flashes the lamps alternately and repeatedly if both of its inputs are at 6 V, but not if one or both is at 0 V.

(ii) How would you arrange for the periods for which the two lamps in (i) are on to be in the ratio 4 :1? [H]

Q MAGNETIC FIELDS AND CURRENTS

Q1 The turns for a solenoid are wound to fill the space between two concentric cylinders of fixed radii. Show that, for a given magnetic flux density along the axis of the solenoid, the heat dissipated in the windings is independent of the diameter d of the wire used. [H]

Q2

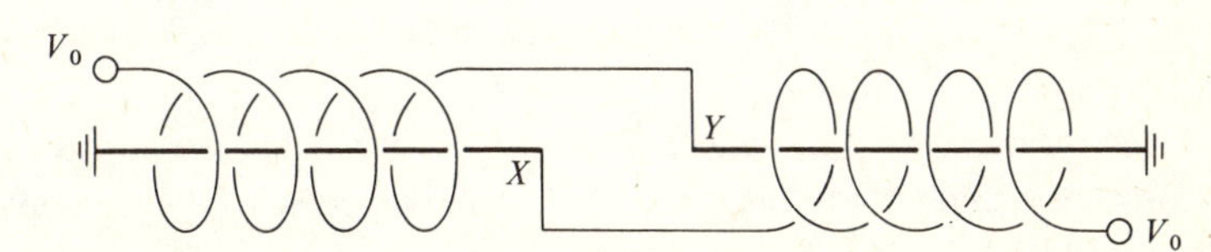

The diagram shows a device containing two identical interconnected units. Each unit consists of a long coil of resistance 2 Ω, along the axis of which runs a superconducting wire which has zero resistance when in a superconducting state, but has a resistance of 10 Ω when in its normal state. A magnetic flux density $B_0 = 10^{-2}$ T will drive either wire from the superconducting to the normal state.

(i) Show that when connected as shown the device behaves as a bistable logic element.

(ii) If the coils have $n = 10^4$ turns per metre, within what limits should the voltage V_0 lie? [H]

Q3 (i) Starting from the fact that, in magnitude, the contribution to the magnetic flux density at a point P from a current i flowing in a length dx of a conductor distant r from P is $(\mu_0 i/4\pi r^2)\,\mathrm{d}x$, show that the magnetic flux density at a point Q on the axis of a single-turn circular coil of radius a carrying a current i is

$$\frac{\mu_0 i a^2}{2(a^2 + y^2)^{3/2}},$$

where y is the distance of Q from the plane of the coil.

(ii) Now consider two such coils placed in the planes $x = \pm \frac{1}{2}a$, and with their centres on the x-axis. Their resultant axial magnetic flux density varies with x, the distance from the axial point midway between the coils, as

$$B(x) = \tfrac{1}{2}\mu_0 i a^2 \left(\frac{1}{[a^2 + (\frac{1}{2}a + x)^2]^{3/2}} + \frac{1}{[a^2 + (\frac{1}{2}a - x)^2]^{3/2}} \right).$$

By taking units in which $a = 1$ and evaluating the variation of $B(x)$ with x for $-0.5 < x < 0.5$, show that the magnetic flux density is very nearly constant between the coils (and in particular is uniform to better than 0.2% for $-0.2 < x < 0.2$). This method was used by Helmholtz to produce a uniform flux density over a usable distance. [H (i)]

Q4 (i) Estimate the current which, flowing round the equator, would be needed to cancel the Earth's magnetic field of 6×10^{-5} T at the North Pole.

(ii) In which direction would it need to flow? [H]

Q5 A magnetic field of flux density 0.75 T is applied in the positive z-direction perpendicular to the plane of a current-carrying copper ribbon of width $2a = 2 \times 10^{-2}$ m in the y-direction. The current flows in the positive x-direction.

(i) If a voltage of 1.5 μV is generated across the width of the ribbon, deduce the drift velocity of the electrons, assuming that they are the only carriers.

(ii) Which edge of the ribbon is at a higher potential?

Q6 Particles of mass m carrying the same charge e as an electron carries are projected horizontally with speed v into a region in which there is a uniform vertical magnetic field of flux density B.

(i) Ignoring gravity, obtain expressions for the radius and period of the circular orbits they describe.

(ii) The same results as in (i) hold in the theory of special relativity provided m is replaced by $m_0(1 - v^2/c^2)^{-1/2}$, where m_0 is called the 'rest mass' of the particle. Find the rest mass of a π-meson (charge e) which completes an orbit of radius 0.25 m in a time of 7.0×10^{-9} s when the magnetic flux density is 2 T. [H (ii)]

Q7 Electrons of a particular speed $v (\ll c)$ are injected at a given point P into a uniform magnetic field.

(i) Show that, provided the various initial angles θ made with the field direction are small, the trajectories will essentially all pass through one point Q some distance L downstream.

(ii) Calculate this distance if $v = 10^7\,\mathrm{m\,s^{-1}}$ and the magnetic flux density B is 10^{-4} T. [H]

Q8 A long straight wire and a wire rectangle of sides 200 mm and 100 mm are arranged in the same vertical plane and such that the longer sides of the rectangle are parallel to the straight wire, the nearer one being 50 mm from it. A current of 2 A flows down the wire and a current of 3 A around the rectangle, the sense of the latter being such that the wire and the side of the rectangle nearest it carry parallel currents.

(i) What is the resultant force on the rectangle?

(ii) What forces, if any, due to the current in the straight wire, act on one of the shorter sides of the rectangle? [H; H (ii)]

Q9 Direct current (d.c.) power is supplied at 33 kV to a distant load by means of two parallel wires having a capacitance of $10^{-11}\,\mathrm{F\,m^{-1}}$ between them. At what (non-zero) electrical load level is there no total force between the wires? For two parallel thin wires, separated by a distance d and each carrying a charge σ per unit length, the force per unit length, F, on each wire is given by $F = \sigma^2/2\pi\epsilon_0 d$. [H]

Q10 A small square loop of side a is suspended with its sides vertical at the centre of a large circular loop of radius R by a fibre which exerts a restoring couple $k\theta$, where θ is the angle between the planes of the loops, and k is a constant. A current i flows in the small loop and a current I flows in the opposite sense in the large loop. (i) Find the minimum value k_0 of k necessary to ensure that the position $\theta = 0$ is one of stable equilibrium. What happens if (ii) $k = (2/\pi)k_0$, and (iii) $k = \frac{1}{2}k_0$? [H]

Q11 A small circular loop of thin wire of mass m is pivoted about a vertical axis at a distance r from a long vertical wire in the plane of the loop, r being large compared to the radius a of the loop. A current i flows in the loop, downwards in the part of the loop closest to the wire. What happens if a upward current I in the wire is switched on? Ignore the Earth's magnetic field. The moment of inertia of the loop about a diameter is $\frac{1}{2}ma^2$.

Q12 (i) The diagram shows a section across the diameters of two identical thin circular conducting rings A and B, each carrying a current I. The rings, of radius r, lie in the planes $z = \pm h$ and both have their centres on the z-axis. In order for the two rings to repel each other, should the two currents flow in the same or opposite directions?

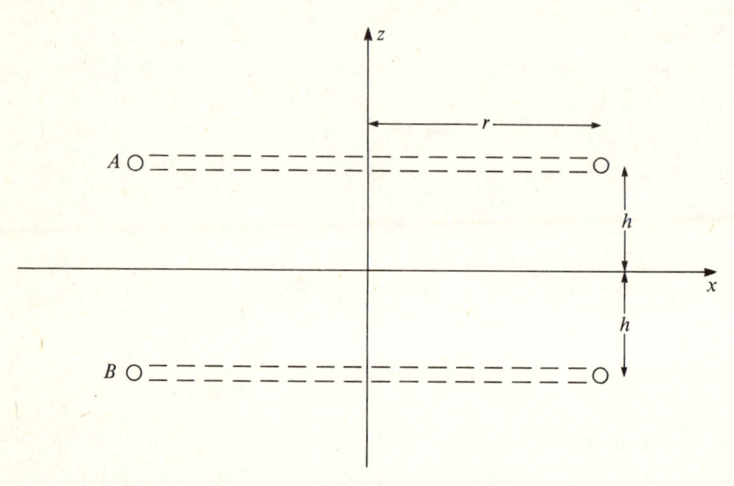

(ii) A current-carrying ring can be floated above a horizontal superconducting plane without any mechanical support. If A represents such a ring and the plane $z = 0$ represents the surface of the superconductor, then in the region $z \geqslant 0$ the magnetic flux density is the same as that in the case considered in (i). If A has mass M and $r \gg h$, show that the equilibrium height h_0 of A above the superconducting surface is given by

$$h_0 = \frac{\mu_0 r I^2}{2Mg}$$

(iii) If the floating ring is displaced slightly in the vertical direction show that the period of subsequent small oscillations is $2\pi(h_0/g)^{1/2}$. [H (ii), (iii)]

Q13* Two identical small magnets of moment M are glued to opposite ends of a wooden rod of length L, one, X, parallel to the rod and the other, Y, perpendicular to it.

(i) Calculate the couples which the magnets exert on each other, and show that they are *not* equal and opposite.

(ii) Explain what would happen if the system were freely suspended from its centre of gravity. Ignore the Earth's field, but show quantitatively how the rod's behaviour comes about. [H; H (ii)]

R ELECTROMAGNETISM

R1 A closed tube of rectangular cross-section measures 10 mm × 15 mm × 100 mm. It is filled with mercury and placed in a magnetic field of flux density 2 T parallel to the 15 mm dimension. A current of 10^3 A is passed across the tube parallel to the 10 mm dimension. What is the pressure difference between the ends of the tube?

R2 A sensitive electrostatic voltmeter is connected between the banks of the River Thames at Tower Bridge where the river is tidal and 100 m wide. What will be recorded if its readings are continuously monitored? Take the maximum speed of flow as $2\,\mathrm{m\,s^{-1}}$, and the Earth's magnetic field as having a flux density of 5×10^{-5} T and an angle of dip of 66° at London. [H]

R3 The wing span of a jumbo jet is 80 m, its length 60 m, and its depth 8 m. Estimate the electrostatic potential differences that could be detected over the surface of the jet when it flies horizontally at $720\,\mathrm{km\,h^{-1}}$:

(i) over the North Pole,

(ii) northwards over the equator,

(iii) eastwards along the equator,

(iv) northwest over London.

The Earth's magnetic flux density is 3×10^{-5} T at the equator, 5×10^{-5} T over London and 6×10^{-5} T at the North Pole. Its angle of dip at London is 66°.

R4 A copper rod of length L is pivoted at a distance x from one of its ends and rotates about an axis perpendicular to its length with constant angular frequency ω. A uniform flux density B is applied parallel to the axis. What e.m.f. is developed between the ends of the rod? [H]

R5 A generator consists of a metal disc of radius 1.2 m and thickness 10 mm, supported by a vertical axle of radius 20 mm. It is rotated in a vertical uniform magnetic flux density of 10 T at a frequency of 5 Hz. Fixed leads are connected to the axle and to the rim of the disc by sliding metal contacts.

(i) Calculate the e.m.f. developed in the generator.

(ii) If the resistivity of the metal is $10^{-7}\,\Omega$ m and the axle has negligible resistance, calculate the total current flowing in the axle when the circumference of the disc is uniformly short-circuited to the axle. [H (ii)]

R6* Two brass gear wheels of radii 10 mm and 50 mm are mounted, in mesh, on shafts of radii 1 mm in a metal frame and are placed in a uniform horizontal magnetic flux density of 0.5 T parallel to the shafts. A mass of 10^{-2} kg is supported by a light thread wound round the shaft of the larger gear wheel. The total electrical resistance between the points of contact of the wheels, when they are just separated, is $10^{-2}\,\Omega$, and frictional losses can be ignored. Find the terminal speed of the mass when falling under gravity. [H]

R7 A long solenoid of radius a with n turns per unit length is wound from wire of radius r and resistivity ρ. Show that, if its ends are shorted, any current initially flowing in it drops to e^{-1} of its original value in a time $\mu_0 \pi n r^2 a/2\rho$. [H]

R8 A device for measuring magnetic fields consists of a small circular coil of 20 turns of very fine copper wire wound on a spool 10 mm in diameter. The coil is inserted into the field and made to spin at a fixed frequency of 50 Hz about a diameter perpendicular to the field.

(i) If the alternating e.m.f. in the coil is found to be 0.5 V r.m.s., calculate the magnetic flux density.

(ii) If the resistance of the coil is $2\,\Omega$ and its ends are short-circuited, find the mean torque exerted by the field on the coil.

(iii) What happens to the current if the number of turns n is increased from 20 to a large value? [H (ii), (iii)]

R9 A bicycle dynamo consists of a small permanent bar magnet which is fixed at its centre to the axle of one of the bicycle wheels, which itself has radius r. Flux from the magnet links (via an iron yoke) a coil of self-inductance L, the coil being connected to a lamp of resistance R. The flux Φ linking the coil can be approximated as a sinusoidally varying quantity of the form $\Phi_0 \cos(\omega t)$.

(i) Establish the equations satisfied by I_0 and ϵ, if the current in the circuit is $I_0 \cos(\omega t + \epsilon)$, and show that $\tan \epsilon = Rr/vL$, where v is the road-speed of the bicycle.

(ii) Find how the power P delivered to the lamp varies with v, and show that it cannot exceed $R\Phi_0^2/2L^2$. [H (i), (ii)]

R10 Charged particles moving in a circular orbit may be accelerated by the e.m.f. which arises when the magnetic flux enclosed by the orbit changes. Show that, if the radius of the orbit remains unchanged, the magnetic flux density at the orbit must be one-half of the average flux density within the orbit. Take the magnetic field as perpendicular to the plane of the orbit. [H]

R11* A uniform thin wire of length $2\pi a$ and resistance r has its ends joined to form a circle. A small voltmeter V of resistance R is connected by leads of negligible resistance to two points on the circumference of the circle at angular separation θ as shown in the two figures.

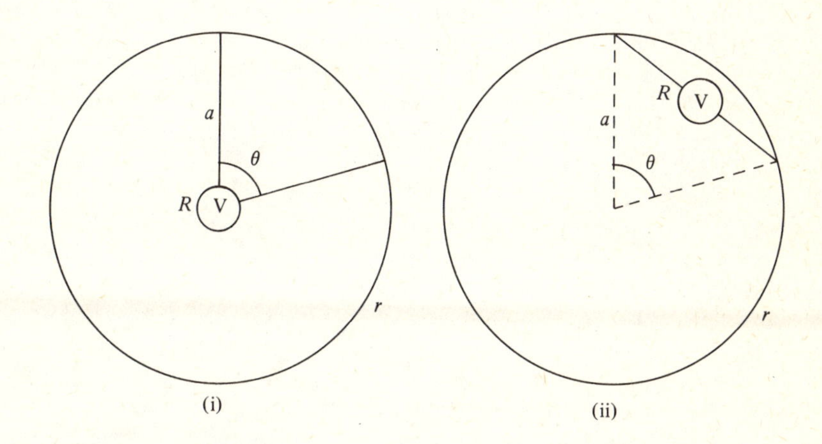

(i) (ii)

A uniform magnetic flux density perpendicular to the plane of the circle is changing at a rate $\dot{B}$. What will the voltmeter read in the two cases? [H (i), (ii)]

R12* A stream of mercury of resistivity ρ fills a horizontal rectangular pipe of length l. Two opposite walls of the pipe, a distance a apart, and of height b, are conducting, the other two walls are insulating. A magnetic field of flux density B is applied parallel to the conducting walls and perpendicular to the length of the pipe and the conducting walls are shorted. If the speed v of the mercury along the length of the pipe is proportional to the total external force acting upon it, show that, when a pressure difference P is applied across the length of the pipe, the speed is

$$v = v_0 \left(1 + \frac{v_0 l B^2}{P\rho}\right)^{-1},$$

where v_0 is the speed at a pressure difference P in the absence of a magnetic field. [H]

R13* Two long wires are placed on a pair of parallel rails perpendicular to the wires. The spacing d between the rails is large compared with x, the distance between the wires. Both wires and rails are made of a material of resistance ρ per unit length. A magnetic flux density B is applied perpendicular to the rectangle so formed, and one wire is moved along the rails with a uniform speed v while the other is held stationary. Determine how the force on the stationary wire varies with x and show that it vanishes for a value of x approximately equal to $\mu_0 v/4\pi\rho$. [H]

S NUCLEI

S1 (*a*) Protons are accelerated from rest through a p.d. of 2×10^6 V and then allowed to fall on a gold ($^{197}_{79}$Au) foil. Find the distance of closest approach of a proton to a gold nucleus.

(*b*) At about what incident deuteron energy would the scattering of deuterons by gold nuclei cease to follow the Rutherford scattering law? The density of nuclear matter is 2.3×10^{17} kg m^{-3}. [H *a, b*]

S2 (*a*) The luminous dials of watches are usually made by mixing a zinc sulphide phosphor with an α-particle emitter. What mass of radium (mass number 226, half-life 1620 years) is needed to produce an average of ten α-particles per second for this purpose?

(*b*) A radioactive element with a half-life of 100 days emits β-particles of average kinetic energy 8×10^{-14} J which are absorbed in a device which converts the energy into electrical energy with an efficiency of 5%. Calculate the number of moles of the element required to generate electricity at an initial power of 5 W. [H]

S3 A radioactive source contains a mixture of two unrelated radioactive substances, A and B, A having a larger decay constant than B. A counter, which is 60% efficient at detecting all decays of nuclei of substance A, but only 11% efficient for those of substance B, gives the following results:

Time/days	0.5	1.0	2.0	3.0	4.0	5.0	7.0	9.0	12.0	15.0
Counts per minute	7000	3000	620	260	180	142	104	76	46	28

Estimate the two half-lives and the numbers of nuclei of each type initially present. [H]

S4 A nucleus X decays, with decay constant λ_X, into a daughter nucleus Y. The daughter nucleus is also radioactive with decay constant λ_Y. At time $t = 0$

there are no daughter nuclei and N_0 nuclei X. Verify, by substitution into the relevant equation and by demonstrating that it has the correct initial value, that the number of daughter nuclei $n(t)$ after time t is given by

$$n(t) = \frac{N_0 \lambda_X}{\lambda_Y - \lambda_X} [\exp(-\lambda_X t) - \exp(-\lambda_Y t)].$$

[H]

S5 In a reactor and separation-plant complex, element A is produced in pure form, without any contamination by the element B into which it decays by β-decay with a half-life of 23 minutes. Element B decays with a half-life of 23 days by β- and γ-ray emission to a stable element C.

A physics professor took delivery of a radioactive sample containing the elements A, B and C produced in a reactor 100 miles away. The sample came with a note saying that $11\frac{1}{2}$ minutes after production of the sample its γ-ray count rate was $1000\ s^{-1}$. The professor found that the count rate of the sample on arrival was also $1000\ s^{-1}$. Had he reason to complain about the transit time T?

[H]

S6 Part of the radioactive series which starts with $^{238}_{92}U$ and ends with $^{206}_{82}Pb$ is given below, together with the half-life of each isotope.

$^{238}_{92}U$	$\rightarrow \ldots \rightarrow$ $^{234}_{92}U$	$\rightarrow$ $^{230}_{90}Th$	$\rightarrow$ $^{226}_{88}Ra$	$\rightarrow$ $^{222}_{86}Rn$ $\rightarrow$
4.5×10^9 y.	2.5×10^5 y.	7.5×10^4 y.	1620 y.	3.83 d.

$^{218}_{84}Po$	$\rightarrow \ldots \rightarrow$ $^{206}_{82}Pb$
3.05 m.	stable

(i) A sample of ore contains all the isotopes in the above series in radioactive equilibrium. If there is 1 kg of $^{238}_{92}U$ in the sample, how many atoms of $^{226}_{88}Ra$ would you expect to find?

(ii) Chemical methods are used to extract in pure form all the $^{226}_{88}Ra$ which is then placed in a sealed container. Make a sketch of how the number of $^{222}_{86}Rn$ atoms in the container will vary during the next 20 days.

(iii) If the abundances of $^{238}_{92}U$ and $^{206}_{82}Pb$ in a sample of ore are A_U and A_{Pb}, what is the approximate age of the Earth? [H (i)]

S7* Part of the series of isotopes produced by the decay of thorium-232, together with their half-lives, is given below.

$^{232}_{90}Th$	$\rightarrow$ $^{228}_{88}Ra$	$\rightarrow$ $^{228}_{89}Ac$	$\rightarrow$ $^{228}_{90}Th$	$\rightarrow$ $^{224}_{88}Ra$	$\rightarrow$ $^{220}_{86}Rn$
1.41×10^{10} y.	5.7 y.	6.13 h.	1.91 y.	3.64 d.	56 s.

Thorium-232 and thorium-228 in equilibrium are extracted from an ore and purified by a chemical process. Estimate how many atoms of radon-220 you would expect to find in 10^{-3} kg of this material after (i) 2 months, and (ii) 200 years. (iii) Sketch the form of the variation in the amount of radon-220 present over a (logarithmic) range from 10^{-3} years to 10^3 years. [H (i), (ii)]

S8 The Q values of a number of nuclear reactions are as follows:

(*a*) $^{31}P(\gamma, n)^{30}P$ −12.313 MeV
(*b*) $^{31}P(p, \alpha)^{28}Si$ 1.916 MeV
(*c*) $^{28}Si(d, p)^{29}Si$ 6.251 MeV
(*d*) $p(n, \gamma)d$ 2.226 MeV
(*e*) $d(d, \gamma)\alpha$ 23.847 MeV

Deduce the Q values for (i) $^{30}P(d, p)^{31}P$, (ii) $^{28}Si(\alpha, n)^{31}P$, and (iii) $^{29}Si(d, n)^{30}P$. [H (i), (ii), (iii)]

S9 The Q value for the reaction $^{19}_{9}F(\alpha, p)^{22}_{10}Ne$ is 1.675 MeV and that for $^{22}_{10}Ne(n, \alpha)^{19}_{8}O$ is −5.711 MeV. What is the minimum energy neutron which, incident upon a carbon tetrafluoride target, can induce the reaction $^{19}_{9}F(n, p)^{19}_{8}O$? [H]

S10 It is believed that one of the processes taking place inside suitably hot second-generation red giant stars is as follows:

(1) $^{12}_{6}C$ captures a proton to form a nucleus A and a γ-ray (radiative proton capture).

(2) A undergoes β^+-decay to produce B.

(3) B transforms to D by radiative proton capture.

(4) D also undergoes radiative proton capture to produce nucleus E.

(5) E transforms to F through β^+-decay.

(6) F captures a proton to yield
either (*a*) G and an α-particle,
or (*b*) H and a γ-ray.

In case (*b*)

(7) H captures a proton radiatively to form I.

(8) I decays by β^+-emission to J.

(9) J captures a proton to yield K and an α-particle.

(i) Identify the ten nuclei $A, B, D, \ldots K$, showing in particular that D and K are the same.

(ii) Establish that two closed cycles are possible on the basis of the reactions given. What are the basic reactions for the two cycles, and which nucleides behave merely as 'catalysts'?

(iii) How much energy is released in converting 1 g of hydrogen to its final product? Use the data given in question **S8** and the fact that the Q value for the decay $p \rightarrow n + e^+ + \nu$ is -1.805 MeV. [H (iii)]

S11* The unstable isotope ^{107}Cd decays to a particular excited state of ^{107}Ag through three different mechanisms, (i) direct (K-)electron capture, (ii) K-electron capture to a higher excited state of ^{107}Ag followed by the emission of a γ-ray photon of energy 0.846 MeV, and (iii) positron (β^+) emission, the maximum kinetic energy of the positron being 0.320 MeV. Find the recoil energies of the nuclei in the two electron capture processes. [H]

S12 On a certain simple model of the nucleus, in which protons and neutrons are not differentiated, the binding energy B for a nucleus of mass number A is given by

$$B(A) = \frac{\alpha A^2}{\beta + A^{1.1}},$$

where $\alpha = 15.3$ MeV and $\beta = 12.4$. On the basis of this model, determine:

(i) The mass number A_s and binding energy (in joules) of the most stable nucleus.

(ii) The mass number A_f of the lightest nucleus capable of undergoing spontaneous fission into two equal parts.

(iii) Whether a formula of the given form, but with different values of α and β, could be made to predict the experimental values which are (approximately) $A_s = 60$ and $A_f = 230$?

(iv) Whether fusion of two identical nuclei is possible. [H (i), (ii)]

S13 A liquid drop model of the nucleus suggests the following (simplified Weizsacker) formula for the mass M of a nucleus of mass number A and atomic number Z:

$$M = Zm_p + (A - Z)m_n - a_1 A + a_2 A^{2/3} + a_3 \frac{(A - 2Z)^2}{A} + a_4 \frac{Z^2}{A^{1/3}},$$

where m_p and m_n are the masses of the proton and neutron respectively and the a_i are positive constants.

(i) Show that the most stable nucleide of mass number A has atomic number Z given by

$$\frac{Z}{A} = \frac{m_\mathrm{n} - m_\mathrm{p} + 4a_3}{8a_3 + 2a_4 A^{2/3}}. \qquad (\dagger)$$

(ii)* Assume now that the result (†) can be approximated as $Z = \lambda A$ ($\lambda \approx 0.46$). Show that the model predicts that, since iron is the element with the largest binding energy per particle, spontaneous fission into two equal fragments should occur for elements with mass numbers higher than that of bromine. [H (ii)]

T ELECTRONS, PHOTONS AND ATOMS

T1 A beam of electrons passes between two parallel conducting plates a distance 10 mm apart, the dimensions of the plates being large compared to their separation. A p.d. of 180 V is applied across the plates and a magnetic field of flux density 5×10^{-3} T is applied perpendicular to the directions of the beam and the electric field. If the electrons which emerge from the plates continue in the same uniform magnetic field outside the plates, what is the radius of curvature of their path? [H]

T2 A 'Millikan-type' experiment is used to try to detect the existence of new and unusual particles called 'quarks'. Oil is sprayed into the apparatus and the voltage V_1 needed to hold a particular drop at rest in the field is noted. The sample believed to contain quarks is then introduced and, if the drop later starts to move, the voltage V_2 needed to restore it to rest is recorded. Below are the entries for six drops. What can you conclude about the likely charge of quarks? The mass of a quark is $\ll 10^{-15}$ kg.

Case	Mass of drop/10^{-15} kg	V_1/V	V_2/V
a	3	459	690
b	1	306	230
c	3	307	230
d	2	612	1836
e	2	305	204
f	4	613	525

[H]

T3 When light is shone on a clean gold surface in vacuo photoelectrons are emitted. Experimental values of the wavelength λ of the light and of V_s, the p.d. which must be applied between the gold and a collecting plate to suppress the photocurrent, are

$\lambda/10^{-9}$ m	216	184	160	142
V_s/V	1.00	2.00	3.00	4.00

Assuming standard values for the other constants involved, deduce values for (i) the Planck constant, and (ii) the work function ϕ_0 of gold. [H]

T4 Antiparticles have the same masses as the corresponding particles and can annihilate them to produce two γ-rays. Calculate the wavelength λ of the γ-rays produced when (i) an electron and a positron, and (ii) a proton and an antiproton, annihilate each other. [H]

T5 A sodium vapour lamp ($\lambda \approx 6 \times 10^{-7}$ m) radiates 100 W of output uniformly in all directions. At what distance from the lamp will the photons have an average density of $10^6\ \mathrm{m}^{-3}$? [H]

T6 The ionization potential $\mathcal{E}$ of atomic hydrogen is 13.58 V. Ultraviolet light of wavelength 8.5×10^{-8} m is incident upon hydrogen gas. Ignoring the thermal energy and molecular binding of the atoms, find the possible kinetic energies T_n with which electrons will be ejected from the hydrogen. [H]

T7 An unusual form of atom is the so-called π-mesic atom. In such an atom a π-meson, which has a mass of 2.4×10^{-28} kg and the same charge as the electron, is in a circular orbit of radius r about the nucleus with an orbital angular momentum $h/2\pi$.

(i) If the radius of a nucleus of atomic number Z is given by $R = 1.7 \times 10^{-15} Z^{1/3}$ m, estimate a limit on Z for which π-mesic atoms might exist.

(ii) Is it an upper or a lower limit? [H]

T8 In a Franck–Hertz type experiment using sodium gas, it was found that the current through the apparatus showed reductions at tube voltages of 2.10 V, 3.18 V, 3.75 V and 4.34 V. When sodium gas is heated and its emission spectrum examined, lines are found at the following wavelengths; 2.853×10^{-7} m, 3.303×10^{-7} m, 5.893×10^{-7} m, 1.139×10^{-6} m, 2.207×10^{-6} m. Reconcile these observations and suggest other wavelengths which should be observed in the emission spectrum. [H]

T9* (i) Light emitted by a hot vapour is analysed using a normal-incidence diffraction grating with 5×10^5 lines m^{-1}. In the line spectrum so formed lines A–E appear at angles to the undiffracted beam $\theta_A, \theta_B, \ldots$, which are found to take the following values:

$$\theta_A = 8.34°, \quad \theta_B = 11.16°, \quad \theta_C = 11.48°, \quad \theta_D = 18.48°, \quad \theta_E = 35.49°.$$

Determine the smallest set of energy levels for an atom of the vapour which will account for these observations.

(ii) There is in the spectrum evidence for a further line (F) at an angle θ_F in the range 24°–25°. Show that this is consistent with the result of (i) (i.e. without addition to the assumed levels).

(iii) At what angle in the range 10°–12° should a further line (G) have been observed? [H (i), (ii)]

T10* The *relativistic* relationship between the *total* energy E of an electron, its momentum p and its rest mass m_e is $E^2 = p^2c^2 + m_e^2c^4$, where c is the speed of light in a vacuum. A photon of frequency ν is scattered, after colliding with an electron initially at rest, through an angle of 90°. Show that its frequency ν' after being scattered is given by

$$\nu' = \frac{m_e c^2}{h\nu + m_e c^2}\,\nu.$$

[H]

T11* An electron of mass m is in a one-dimensional well for which the potential is zero along a length a (taken as $0 \leqslant x \leqslant a$) and infinite outside this length. Using the one-dimensional Schrödinger equation;

(i) Obtain an expression for the energy of the electron when it is in its nth lowest state, $n = 1$ corresponding to the lowest value of the energy.

(ii) Determine where the electron is most likely to be for $n = 1$ and $n = 2$.

(iii) Find the probability P that the electron is within a distance ϵ of $x = 0$.

(iv) Interpret the result of (iii) when n becomes very large. [H (i), (iii)]

T12 A particle of mass m is confined to a rigid two-dimensional box of rectangular cross-section $a \times b$. Assuming the wave function (which obeys the relevant Schrödinger equation) has the form

$$\psi = C \sin\left(\frac{2\pi x}{\lambda_1}\right) \sin\left(\frac{2\pi y}{\lambda_2}\right),$$

find the lowest allowed energy level of the particle. [H]

T13* A particle of mass m moves in a one-dimensional harmonic potential, $V(x) = \frac{1}{2}kx^2$.

(i) Show that $\psi(x) = A \exp(-\alpha x^2)$ is a solution of the appropriate Schrödinger equation provided that α has a particular value.

(ii) Deduce the corresponding value of the energy E of the particle, expressing it in terms of the frequency of oscillation ν of a classical particle of mass m moving under the influence of a spring of spring-constant k. [H (i)]

U SHORTS

U1 By how much would the time of sunrise change if the speed of light were to double overnight?

U2 How would you determine, non-destructively, and with eyes closed, which is which amongst three identically-shaped metal bars known to consist of a permanent magnet, a piece of soft iron and a piece of copper?

U3 A man standing on the South bank of a still river of width w, which runs East-West, can run with speed V_1 and swim with speed $V_2 (< V_1)$. What path should he take to reach a point on the North bank, a distance d to the East, in the shortest possible time? [H]

U4 Hydrogen gas can be driven outwards from a star as a result of absorbing photons emitted by the star. What speed will a hydrogen atom acquire as a result of absorbing a photon emitted by a sodium atom in the star? [H]

U5 The annual cost of transmitting required electrical power a given distance, at a fixed voltage, is the sum of the annual cost P_1 of energy wastage, and the annual interest P_2 on the capital cost of the material of the cables. What is the optimum ratio for P_1/P_2? [H]

U6 An electric motor supplies 0.8 kW to drive a flywheel at a constant rotational speed of 20 rev s^{-1}. If the power is switched off the constant frictional couple at the bearings brings the flywheel to rest in 40 s. What is the moment of inertia of the rotating part of the system? [H]

U7 An ammeter of internal resistance 10 Ω is connected by wires of negligible resistance between the two rails of a (steam) railway near London. What will it register when an express train passes over it?

U8 In the pulley system shown, the pulleys of radii a and b are rigidly fixed together but free to rotate about their fixed common axle. The weight W hangs from the axle of the freely suspended pulley C which may rotate about that axle. If section 1 of the (rough) rope is pulled downwards with velocity V, which way will W move and with what speed?

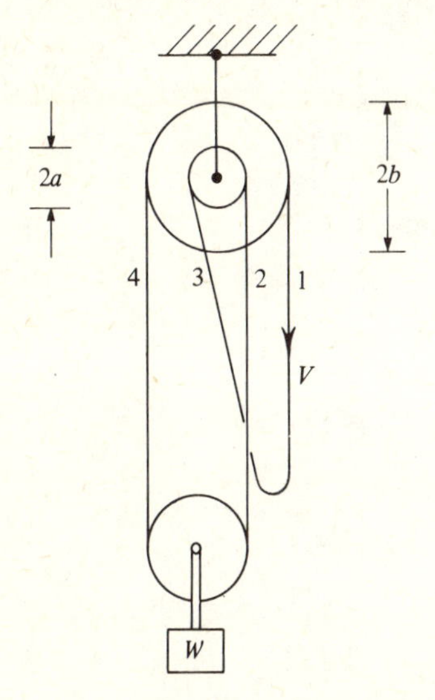

[H]

U9 A uniform cylinder of mass m hangs from a spring balance A and is lowered slowly and steadily into a bowl of water, of mass M, until it rests totally submerged on the bottom of the bowl. The bowl is throughout positioned on the scale-pan of a second spring balance B. Draw the variation of the scale readings of A and B as a function of the vertical position x of the cylinder (measured downwards).

U10 Find the approximate rotation frequency of a nitrogen molecule at 600 K if its kinetic energy of rotation as a rigid dumb-bell is equal to its translational kinetic energy.

U11 Design a simple electronic circuit, based on at most resistors, capacitors and inductors, which will send high frequency signals to one loud speaker and low frequency ones to another.

U12 Determine how the period of (shape) oscillations of a freely-falling drop of mercury depends upon the radius of the drop. [H]

U13 If hen's eggs should be boiled for five minutes, for how long should an ostrich's egg be boiled? [H]

V LONGS

V1 How much brighter is sunlight than moonlight? [H]

V2

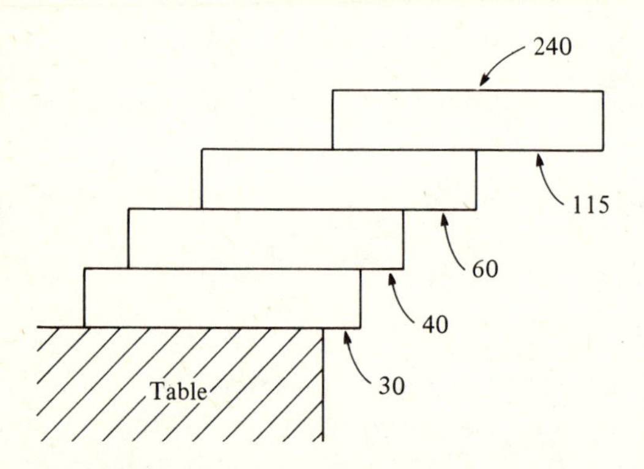

The figure shows a pile of four identical uniform books and the amounts by which each overhangs the one below it (or the table). All dimensions are in mm. Does the pile topple? Justify your answer. [H]

V3

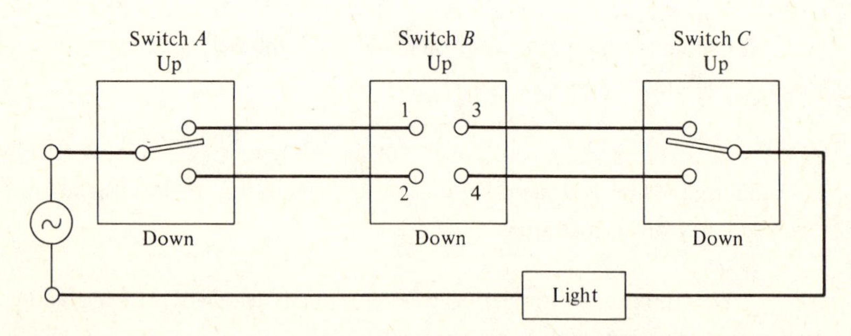

The figure shows a house lighting circuit in which any one of three switches may be used to turn a single light on or off. All three switches have only a single external control that can be either up or down.

(i) How is switch B connected internally?

(ii) Show how five switches can all be made to control one light. [H (ii)]

V4 The level ground in an area in which geophysical explorations are being carried out consists of a uniform layer, of depth h, of a rock in which the velocity of sound is V_1, on top of rock in which it is $V_2 (> V_1)$. Find the minimum distance x apart of an explosion and a microphone, which detects sound waves from the explosion, such that the sound wave which has travelled just inside the lower layer of rock arrives at the microphone before the direct surface wave. [H]

V5 A relay coil has an inductance of 0.5 H and a resistance of 20 Ω. The coil is suddenly connected to a source of constant p.d. of 10 V. After 20 ms the source is disconnected and the terminals of the coil are simultaneously short-circuited. The relay contacts operated by the coil close when the coil current is 0.2 A and rising, and open when it is 0.1 A and falling. For how long do the contacts remain closed? [H]

V6 A hollow ball and a solid ball, each of the same radius r and made of the same material, are released simultaneously from the top of a rough slope (so that both roll without slipping). Air resistance is to be ignored.

(i) Why does one reach the bottom of the slope before the other?

(ii) Do they have equal velocities when they pass a particular point on the slope, even though they do so at different times?

(iii) Could the order of reaching the bottom be changed by choosing a different material for one of the spheres? [H (ii)]

V7 Find the period of vertical oscillations of an iceberg which is in the form of an upright regular pyramid with 10 m of its top showing above the surface. The density of ice is 900 kg m^{-3}. Ignore any motion of the water. [H]

V8 Two identical parallel-plate capacitors are connected in parallel and, after being charged, are isolated.

(i) The space between the plates of one capacitor is now filled with an oil of relative permittivity ϵ_r. By what factor is the electrical energy stored in the system changed?

(ii) The oil is now emptied out. What is the new stored electrical energy? Reconcile your answer with energy conservation. [H]

V9 If Great Britain changed to driving on the right-hand side of the road (instead of on the left), would the length of the day increase, decrease, or remain the same? [H]

V10 A U-tube consists of two vertical limbs joined at their lower ends by a horizontal one of length l, all limbs having the same cross-sectional area. The two vertical limbs contain different liquids of densities ρ_1 and ρ_2. The liquids do not mix but meet at an interface which lies in the horizontal section of the tube at a distance x_0 from the vertical limb containing liquid of density ρ_1. The tube is now rotated about this limb (as vertical axis) with angular speed ω. Show that if the new position of the interface is $x_0 + x$ (still within the horizontal limb), then

$$(\rho_1 - \rho_2)\omega^2(x_0 + x)^2 - 2(\rho_1 + \rho_2)gx + \tfrac{1}{2}\rho_2 l^2 \omega^2 = 0.$$

[H]

V11 Electrons of mass m leaving a vacuum-tube cathode with negligible velocity are attracted towards the anode, which is at a potential V relative to the cathode, and at a distance d from it. The tube is placed between the poles of an electromagnet in such a way that a uniform magnetic field is at right angles to the uniform electric field. If the magnetic flux density exceeds a certain value, B_0, current ceases to flow through the tube. Find an expression for B_0. [H]

V12*

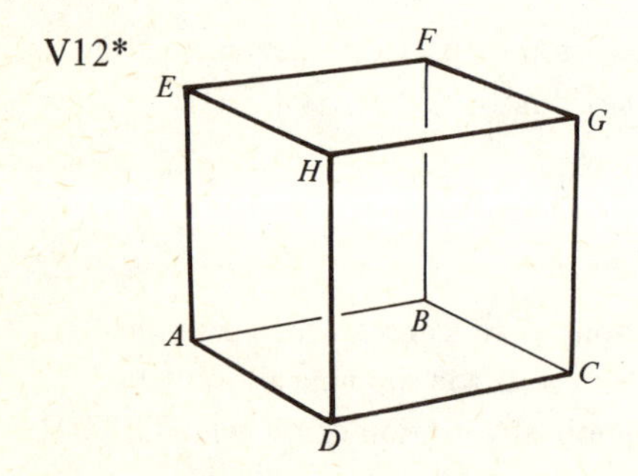

The figure shows twelve wires, each of electrical resistance 1 Ω, formed into a cube. What is the electrical resistance between (i) A and G, (ii) A and H, and (iii) A and D? [H]

V13* A straight rigid uniform hair lies on a smooth table. At each end of the hair sits a flea. Show that if the mass of the hair (M) is not too great relative to that of each of the fleas (m), then by simultaneous jumps with the same speed and angle of take-off the fleas will be able to change ends without colliding in mid-air. [H]

W DATA HANDLING

W1 A group of students measure g in a pendulum experiment and obtain the following values (in $\mathrm{m\,s^{-2}}$): 9.80, 9.84, 9.72, 9.74, 9.87, 9.77, 9.28, 9.86, 9.81, 9.79, 9.82. What would you give as the best value of g for the group, and the standard deviation of this value? Explain your reasoning. [H]

W2 Measurements of a certain quantity gave the following values: 296, 316, 307, 278, 312, 317, 314, 307, 313, 306, 320, 309. Within what limits would you say there is a 50% chance that the correct value lies? [H]

W3 The following are the values obtained by a class of 15 students when measuring a physical quality x: 53.8, 53.1, 56.9, 54.7, 58.2, 54.1, 56.4, 54.8, 57.3, 51.0, 55.1, 55.0, 63.8, 54.2, 56.6.

(i) Display these results as a histogram and state what you would give as the best value for x.

(ii) Without calculation, but explaining your reasoning, give an indication of the amount of reliance that could be placed on your answer to (i).

(iii) Textbooks give the value of x as 53.6 with negligible error. Are the data in conflict with this? [H (i), (ii), (iii)]

W4 The saturated vapour pressure p of water at a temperature θ is related to the saturated vapour pressure p_0 at 100 °C by the empirical expression

$$\theta = 100.00 + 28.012\left(\frac{p}{p_0} - 1\right) - 11.64\left(\frac{p}{p_0} - 1\right)^2 + 7.1\left(\frac{p}{p_0} - 1\right)^3.$$

The boiling points of water at the top and bottom of a building are 99.10 °C and 99.80 °C respectively. Use the above formula, making suitable approximations where possible, to obtain an estimate (accurate to about 1%) of the height of the building, given that the density of air is $1.23\ \mathrm{kg\,m^{-3}}$ and of mercury $1.36 \times 10^4\ \mathrm{kg\,m^{-3}}$. [H]

W5 In an experiment to study the rebound of a hard steel ball from a soft steel anvil, a massive block of mild steel rests on a firm pedestal, its upper surface being smooth and horizontal. A hard steel ball is dropped from various heights h_1 so that it strikes the specimen normally, produces a dent, and rebounds to a height h_2. The permanent indentation is examined with a microscope and its diameter d determined.

If the steel ball, of mass m and Young modulus E_1, is not permanently deformed during impact, and E_2 is the Young modulus of the anvil, it can be shown that

$$h_2 = \frac{d^3P^2}{3mg}\left(\frac{1}{E_1} + \frac{1}{E_2}\right),$$

where P, which has the same dimensions as E_1 and E_2, is a plastic property of the anvil known as its 'yield pressure'.

The accuracy of measurement of h_1 was high, of h_2 1%–3%, and of d 0.01 mm. For the apparatus used $m = 4 \times 10^{-3}$ kg, $E_1 = 3 \times 10^{11}$ N m^{-2} and $E_2 = 1.5 \times 10^{11}$ N m^{-2}. The experimental data was as follows:

h_1/mm	h_2/mm	d/mm
1000	283	1.22
700	211	1.08
500	162	1.01
300	103	0.87
200	73	0.77
100	41	0.64
50	21	0.50
30	14	0.43
18	9.0	0.37
6	3.7	0.27
2	1.3	

(i) How consistent with the theory are the data?

(ii) Assuming the theory is correct, what is the best value for P?

(iii) Plot the ratio of the velocity of rebound to the velocity of impact (the coefficient of restitution) as a function of the velocity of impact. Comment on the result. [H (i), (iii)]

W6 Two physical quantities x and y are connected by the equation

$$y^{1/2} = \frac{x}{ax^{1/2} + b},$$

and measured pairs of values for x and y are:

x	10	12	16	20
y	409	196	114	94

Determine, by graphical means, the best values for a and b. [H]

W7 Measured quantities x and y are known to be connected by a formula of the form

$$y = \frac{ax}{b + x^2},$$

where a and b are constants. Pairs of values obtained experimentally are:

x	2.0	3.0	4.0	5.0	6.0
y	0.32	0.29	0.25	0.21	0.18

Use these data to make the best estimates you can for the values of y which would be obtained for (i) $x = 1.0$, (ii) $x = -3.5$, and (iii) $x = 12.0$. (iv) Rank your three estimates according to their likely accuracy. Indicate the reasons for your ranking. [H]

W8 The variation of a physical quantity $y(T)$ with absolute temperature T is measured and the following results obtained.

T/K	10	20	30	50	100	200
y	0.172	0.719	1.16	1.69	2.25	2.60

(i) Plot $\ln y$ against $1/T$ and determine as far as possible the form of $y(T)$.

(ii) Estimate $y(5)$ and $y(500)$.

(iii) For large T, y can be approximated by $y(T) = a + bT^{-1}$. Determine values for a and b. [H (i), (iii)]

W9 According to a particular theory T_1, two dimensionless quantities X and Y have equal values. Nine measurements of X and seven of Y gave the following results:

X:	22	11	19	19	14	27	8	24	18
Y:	11	14	17	14	19	16	14		

(i) Assuming that the measurements of both quantities have Normal distributions about their true values, determine whether or not they support T_1.

(ii) An alternative theory T_2 predicts that $Y^2 = \pi^2 X$. Treating the errors in the mean values of X and Y as small, comment on whether this theory is better supported by the data. [H]

W10 During an investigation into possible links between mathematics and classical music, pupils at a school were asked whether they had preferences (*a*) between mathematics and English, and (*b*) between classical and pop music. The results are given below.

	Prefer classical music	No preference	Prefer pop music
Prefer mathematics	23	13	14
No preference	17	17	36
Prefer English	30	10	40

(i) Is there any evidence of a correlation between academic and musical tastes?

(ii) Is there any evidence for a claim that the pupils either had preferences in both areas or had no preferences? [H (i), (ii)]

W11 Three candidates X, Y, and Z were standing for election to the students' seat on a School Board of Governors. The electorate (the Sixth Form, consisting of 150 boys and 105 girls) were each allowed to cross out the name of the candidate they least wished to be elected, the other two candidates then being credited with one vote each.

(i) X received 100 votes from boys, whilst Y received 65 votes from girls.

(ii) Z received 5 more votes from boys than X received from girls.

(iii) The total votes cast for X and Y were equal.

Analyse these data in such a way that a χ^2-test can be used to determine whether or not voting was other than random amongst (*a*) boys, (*b*) girls, and (*c*) the electorate at large. [H]

W12 An archer is thought to have a constant probability, for each arrow fired, of hitting the gold part of the target. He fires 100 groups of 5 arrows each and the distribution of the number of gold hits in each group is given below.

Gold hits in group	0	1	2	3	4	5
Number of groups	15	37	29	15	3	1

(i) Do the data support the assumed constancy?

(ii) What is the chance, if he fires a further 100 arrows, that 40 or more of them will hit the gold?

Some time later, after paying for instruction from a Master Bowman, he fires 300 further arrows and records 108 gold hits.

(iii) Did he waste his money?

(iv) How many groups of 5 arrows is he now likely to have to fire before he obtains a group of 5 'golds'? [H (i), (ii), (iii)]

W13 On a certain racecourse there are 12 fences to be jumped and any horse which falls is out of the race. In a season of racing a total of 500 horses started the course and the following numbers fell at each fence:

Fence	1	2	3	4	5	6	7	8	9	10	11	12
No. falling	62	75	49	29	33	25	30	17	19	11	15	12

Determine the overall probability of a horse falling at a fence, assuming the probability to be the same for all horses and fences. Investigate the validity of this assumption and discuss any significant disagreements with the data. [H]

X GUESTIMATION

In this section you are asked to estimate (and not really to guess) the quantities indicated, basing your estimates on approximate theories of the physics, standard constants and reasonable guesses for the basic quantities involved. Usually one significant figure of accuracy is sufficient, and sometimes only the correct order of magnitude can be expected.

It is unlikely that your estimate and the one given in the answers will agree, but you should try to satisfy yourself that the discrepancy is understandable in terms of the different values assumed for basic quantities or of a different physical approach to the estimation.

X1 (*a*) The number of mercury atoms in an ordinary thermometer.

(*b*) The mass of the air you breathe in a year.

(*c*) The impulse due to a raindrop when it hits the ground.

X2 (*a*) The mass of water in all the oceans.

(*b*) The gain or loss in a week of a pendulum clock when it is moved from a warm house to a cold furniture store.

(*c*) The minimum weekly cost of subsistence for a man who sleeps 8 hours per day, if his 'daily diet' could be supplied from the electricity mains and he uses about 75 W of power whilst sleeping, and twice that amount whilst just sitting.

X3 (*a*) The velocity of an atmospheric nitrogen molecule.

(*b*) The power needed to operate a passenger lift.

(*c*) The mass of water in a soap bubble.

X4 (*a*) The maximum theoretical distance at which a car's headlights could be resolved by the naked eye.

(*b*) The minimum volume of a hot air balloon capable of lifting one man.

(*c*) the area of a bicycle tyre in contact with the road when the bicycle is being ridden by a man of mass 70 kg.

X5 (*a*) The mass of cold water that could be brought to the boil using the energy dissipated when a motor car is brought to rest from $100\,\mathrm{km\,h^{-1}}$.

(*b*) The total number of photons emitted by a 100 W light bulb during its lifetime.

(*c*) The ratio of secondary to primary turns of a transformer which, when driven from the a.c. mains, could cause a car spark-plug to fire.

X6 (*a*) The average distance between the excess electrons on one plate of a parallel-plate capacitor whose plates are separated by 1 mm of air and have a potential difference between them of 100 V.

(*b*) The temperature at which the Earth would immediately lose most of its atmosphere.

(*c*) The voltage needed for an X-ray tube to be used in an investigation of the structure of a crystal in which the plane spacing is about $3 \times 10^{-10}\,\mathrm{m}$.

X7 (*a*) The tension in a violin string.

(*b*) The temperature rise of the Earth if all its rotational energy were suddenly converted to heat.

(*c*) The focal length of a (full) goldfish bowl.

X8 (*a*) The angle at which the first principal maximum would be seen if a piece of a long playing record ($33\frac{1}{3}$ r.p.m.) were used as a diffraction grating for white light at normal incidence.

(*b*) The minimum work done in building the Great Pyramid.

(*c*) The change in the temperature of the Earth if it were 10% nearer the Sun.

X9 (*a*) The power produced by a grasshopper when hopping.

(*b*) The number of collisions an air molecule makes per second.

(*c*) The maximum magnetic energy stored in the field of a 1 kW bar fire.

X10 (*a*) The site area needed for a terrestrial 1000 MW solar power station, given that the Sun radiates at a power of about 4×10^{26} W.

(*b*) The power required to keep the air in a house warm in winter, if draughts cause the complete replacement each hour of the warm air by cold air from outside.

(*c*) The length of a tangle of insulated copper wire which weighs 1.35 kg and has an electrical resistance of 0.70 Ω.

X11 (*a*) The power available in a self-winding watch which obtains its energy from the natural changes in atmospheric pressure.

(*b*) The temperature change when two water drops each of radius 2×10^{-6} m coalesce.

(*c*) The energy released by a lightning flash, given that the potential gradient at the Earth's surface under a thundercloud is 10^4 V m^{-1}.

X12 (*a*) The maximum acceleration of the tip of a gramophone stylus playing a long-playing ($33\frac{1}{3}$ r.p.m.) record of orchestral music.

(*b*) The fraction of water on Earth that is in the atmosphere.

(*c*) The number of car batteries needed to store the same energy as is available from one litre of petrol. A typical car travelling at 50 km h^{-1} needs about 3 kW of useful power.

X13 (*a*) How fast a wheel may be spun before a rubber band stretched round its rim detaches itself. The circumference of the wheel is 110 mm and that of the unstretched band 100 mm. The band, which weighs 4 g, increases in length by 15% when hung as a closed loop over a nail, as a support for a 50 g weight.

(*b*) How long would it be before the Sun, whose diameter subtends about $\frac{1}{2}^\circ$ at the Earth's surface, burned itself out if it consisted of burning coal and continued to radiate at its present power of 4×10^{26} W. The heat of combustion of coal is $3 \times 10^7 \text{ J kg}^{-1}$.

(*c*) How many molecules from Socrate's cup of hemlock are likely to be contained in the next glass of water you drink.

Y WHY OR HOW

In this section you are asked to explain qualitatively why or how something happens, formulating your answers in terms of the physics principles involved. Because it is not possible to produce a 'complete and correct' answer to such questions, solutions are not provided. However, if help is needed, then reference to textbooks on the relevant areas of physics will usually provide the information sought.

Y1 (*a*) Why does a blue cloth look black when viewed in sodium light?

(*b*) Why do peas cook more quickly in a pressure cooker than in an open pan?

(*c*) How would you measure the acceleration of a train, whilst locked inside one of its windowless carriages?

Y2 (*a*) Why, if a metal string is needed on a musical instrument, is a steel one used rather than a copper one?

(*b*) How is it possible for trainee astronauts to get experience of gravity-free conditions in a plane which has a ceiling of only 15 km?

(*c*) How does a car speedometer work?

Y3 (*a*) Why does a well-cut diamond sparkle?

(*b*) How does the temperature of a room affect the pitch of a violin, a flute and a record-player?

(*c*) Why does heat have to be supplied to boil a liquid, even though the molecules which have just left the liquid are no hotter than those in the liquid?

Y4 (*a*) Why do the spoked wheels of a stage-coach in a TV Western film often appear to be rotating backwards, but similar spoked wheels on a farm-cart in the same film seldom do?

(*b*) Why is there a force of attraction between a charged particle and an uncharged isolated copper sphere some distance away, and why does the force increase if the sphere is earthed?

(*c*) Why are fluorescent tubes always run with a large inductor, known as a 'choke', connected in series?

Y5 (*a*) How does an air-pressure paint sprayer pick up paint from a paint container positioned below the nozzle?

(*b*) Why are the aperture markings on a camera lens 2.8, 4, 5.6, 8, 11, 16, 22 and not, say, 1, 2, 3, 4, 5, 6, 7?

(*c*) Why, in a gas, is the speed of sound restricted to the average velocity of the molecules, but in a solid suffers no such restriction?

Y6 (*a*) Why is it much easier to balance a broomstick vertically on one's flat hand than to balance a matchstick?

(*b*) How would the period of a simple pendulum, which had a charged metal bob, change if an earthed metal plate were placed beneath the pendulum?

(*c*) Why, if excited atoms give a spectrum of discrete lines, does a red-hot bar of iron radiate a continuous spectrum?

Y7 (*a*) Why is it more comfortable for a passenger if a car's speed is increased when the car goes round a corner?

(*b*) Why is it often difficult to unroll plastic film intended for covering open food-containers?

(*c*) Why, even though the crystals are all randomly aligned, are X-ray diffraction patterns obtainable from powdered salt?

Y8 (*a*) Why is it possible to safely pinch between damp fingers a candle wick hot enough to give off white light, but not an iron bar which is not even warm enough to emit a dull red glow?

(*b*) How is the following explained? Two unequal pieces of iron are magnetized and, when held a long way apart, are found to repel each other. As they are brought closer together with unchanged orientation, the repulsive force increases, then decreases and becomes attractive.

(*c*) Why must the hole in a pinhole camera be neither very large nor very small, even if the time spent on the exposure is not important?

Y9 (*a*) Why, if a thin teacup is tapped on its rim at a point opposite the handle or at a point 90° from the handle, does the sound produced have a lower pitch than if it is tapped at a point 45° from the handle?

(*b*) How does an ice-skater spin more and more quickly on one spot and why does what happens not violate the law of conservation of energy?

(*c*) How would you convince a sceptic that the motions of the solar system are not controlled by electrostatic forces?

Y10 (*a*) Why do car tyres get warm even though the point of contact of the tyre with the ground is always instantaneously at rest, and so no work is done in moving this point in the direction opposite to the frictional force between the tyre and the road?

(*b*) How does TV 'ghosting' - images up to about one-tenth of the screen width behind the main picture - occur?

(*c*) How can the fact that, when a frictionless 'secretarial' swivel chair is given an initial angular velocity, the seat rises but then comes to a halt, be reconciled with the conservation of angular momentum?

Y11 (*a*) Why is the behaviour of a cork released from below the surface of a bucket of water, the bucket being in a lift, dependent upon whether the cork is released before or after the lift cable breaks?

(*b*) Why does a tennis ball served by a right-handed player often swerve to the left (as seen by the server)?

(*c*) Why are the sunset red, the clouds white, and the sky blue?

Y12 (*a*) Why, if an aluminium ring is placed over a vertical iron rod with a coil carrying an a.c. current wrapped round its lower end, does the ring rise and remain suspended part way up the rod?

(*b*) Why is it that a pendulum swinging on a hinge can lose amplitude and heat the hinge, but no amount of steady heating of the hinge with a flame will cause the pendulum to swing?

(*c*) Why can a ladder be leant at an angle on a rough floor and against a smooth wall, but not on a smooth floor and against a rough wall?

Y13 (*a*) Why does a hard-boiled egg stand on one of its ends when spun rapidly, but lie on its side if spun slowly?

(*b*) Why does air friction result in an increase in the speed of an artificial satellite?

(*c*) How does the movement of a yacht, which has a sail at the front and an electric fan which blows air onto the sail mounted at its stern, change as the size of the sail is increased from nothing to much larger than the area swept by the fan blades?

Z SOPHISTRY

In this section you are asked to examine and criticize any loosely-worded or erroneous arguments used, providing corrected versions where appropriate.

Z1 In one of H. G. Wells' science fiction stories, the body of a character named Pyecraft loses all its weight but retains its original volume. Because he then displaces more air than he weighs, by Archimedes' principle he experiences an upthrust and floats fully clothed to the ceiling.

Z2 Since, by Newton's law, action and reaction are always equal and opposite, it follows that when a horse pulls on a cart with a certain force, the cart always pulls back on the horse with the same force. Consequently the horse and cart cannot move.

Z3 Double glazing reduces heat losses through windows because there is twice as much glass for the heat to penetrate. It follows that it really doesn't matter how large the space between the two panes of glass is, and, in fact, one sheet of glass of twice the normal thickness would do just as well as the conventional two-pane arrangement.

Z4 The law of partial pressures states that the pressures exerted by vapours are additive. Consequently the pressure just above the mercury surface in the open limb of a barometer is the sum of the atmospheric pressure and the pressure of mercury vapour at the relevant temperature. Above the mercury in the closed limb there is only the pressure of the mercury vapour. Thus a straightforward subtraction of the heights in the two limbs gives the atmospheric pressure, without any further correction.

Z5 By Archimedes' principle the upthrust force on a body immersed in a fluid is equal to the mass of fluid displaced. Further, by Newton's law, force is equal to mass times acceleration. So all bodies immersed in fluids will rise with unit acceleration when released.

Z6 Electrons repel each other. Two wires carrying parallel currents attract each other. Therefore currents in wires cannot be due to the motion of electrons.

Z7 A book of mass m rests on the floor of a stationary lift. By Newton's third law, action and reaction are equal and opposite, and so the force R exerted by the floor on the book is equal to mg, the weight of the book. When the lift moves downwards with an acceleration $\frac{1}{3}g$, the force R is equal to $\frac{2}{3}mg$. Since this does not equal the weight of the book, Newton's third law is not obeyed in accelerating systems.

Z8 When crossing a river, one usually finds the fastest current where the water is deepest. This shows that water flows slowest when it is close to a fixed surface containing the water. It therefore also follows that if water in a pipe comes to a constriction, where the pipe wall is close to all parts of the aperture, the water will slow down. Since the water in the pipe flows slowly when its cross-sectional area is smaller and faster where the cross-section is larger, a constricted pipe carrying water will never be able to sustain a constant throughput without developing pockets of air in the flow.

Z9 When a nail is driven into a piece of wood by a hammer, the velocity of the nail immediately after impact can be deduced from the law of conservation of momentum. Therefore, provided they are of the same mass, a rubber-headed hammer and a steel one are just as effective as each other.

Z10 Like magnetic poles repel and unlike poles attract. Therefore, since the Earth's North Pole is closer to England than is the South Pole, if a magnet is floated on a cork in a dish of water in London, it will move towards the North end of the dish.

Z11 The walls of a gas container are made of molecules and these attract the molecules of the gas in just the same way as do other gas molecules. Thus gas molecules moving towards the walls accelerate and the measured pressure is higher than it should be.

Z12 The nucleus of an atom cannot consist of a collection of neutrons and protons. This is clear because (i) negatively charged particles, much lighter than either neutrons or protons, are frequently seen to be emitted by nuclei, (ii) the relative atomic masses of most elements are not integral even though neutrons and protons have very nearly equal masses, and (iii) the protons would repel each other and no such nucleus could be stable.

Z13 Because of the bulge of the Earth at the equator, a given mass weighs more there than at the North Pole.

HINTS AND INTERMEDIATE ANSWERS

A Physical dimensions

A1 Recall that e^2/ϵ_0 has the dimensions of Energy × Length.

A2 Use the fact that Resistance × (Charge)^2 has the same dimensions as Power × (Time)^2.

A3 (ii) Show that ϵ_0/μ_0 has the dimensions Ω^{-2}.
(iii) The dimensions of ψ are irrelevant.

A5 $P \propto F^{3/2}\rho^{-1/2}l^{-1}$.

A6 $\sigma = \mu k^4 h^{-3} c^{-2}$.

A9 Write $\text{m s}^{-1} = c/(3 \times 10^8)$ etc., and construct an expression for 1 kg.

A10 Proceed as in question A9.

A11 Note that m_1/m_2 is a dimensionless variable.

A12 (i) Construct two dimensionless variables, one excluding R and one excluding p.
(ii) For scaling, both dimensionless variables must have unchanged numerical values.

A13 (ii) The wavelength is that corresponding to the head of water that gives an unchanged value of $pA^{1/2}\sigma^{-1}$, i.e. an unchanged value of $\rho ghA^{1/2}\sigma^{-1}$. This leads to a head of water of 0.9 m.
(iii) By interpolation, the corresponding head of water h_W is 0.65 m. The value of $\rho h\sigma^{-1}$ must be the same for both water and liquid, thus giving the required value of h_L.

B Linear mechanics and statics

B1 (i) Show that $\frac{1}{2}\pi r^2 \rho x^2 + mx - mh = 0$, where x is the required depth.

B2 Equate horizontal and vertical forces and take moments about either end of the ladder. $\mu = \frac{2}{3}$.

B3

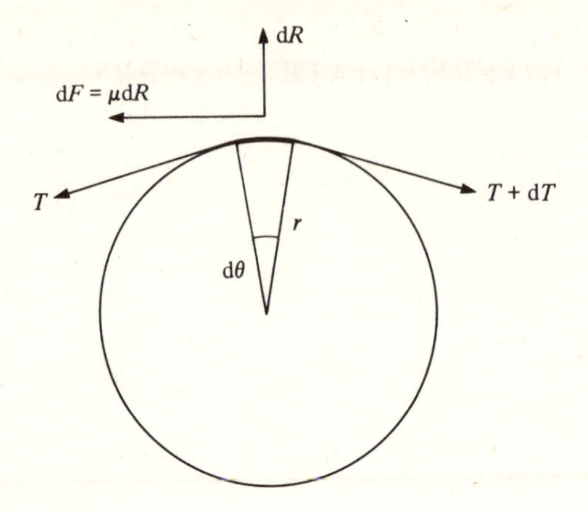

(i) Show that $dR = T d\theta$ where dR is the normal reaction between the bar and a length of rope subtending $d\theta$ at the bar's axis. Establish also that $dT = \mu dR$.

B4 $$\text{Range} = \frac{2V^2 \sin\theta}{g}\left(\frac{EQ}{mg}\sin\theta + \cos\theta\right).$$

B5 (i) Show that $\dot{v} = (F/M)(1 - v^2/U^2)$.
(ii) Write $\dot{v}$ as $v\, dv/ds$.

B6 (ii) Compare couples about the rear edge of the base of the wall. Couple due to the wind $= \frac{1}{2}\rho_A V^2 h^2$. Couple due to gravity $= \frac{1}{2}\rho_W ght^2$. At the critical wind speed, the horizontal force per metre on the wall is 294 N.

B7 (i) Show that $P = A\rho v(V - \alpha V)v$. The rate of K. E. loss $= \frac{1}{2}A\rho v V^2(1 - \alpha^2)$.
(ii) P is optimized when $\alpha = \frac{1}{3}$.

B8 (*a*) Get a bicycle and try it.
(*b*) Consider what is happening to the centre of gravity of (chair + jug + beer).

B9 Show that the two final velocities are $(2\alpha)^{-1}V\sec\theta$ and $(2 - 2\alpha)^{-1}V\sec\theta$, and hence that the released energy is $E[\frac{1}{4}\alpha^{-1}(1 - \alpha)^{-1}\sec^2\theta - 1]$.

B10

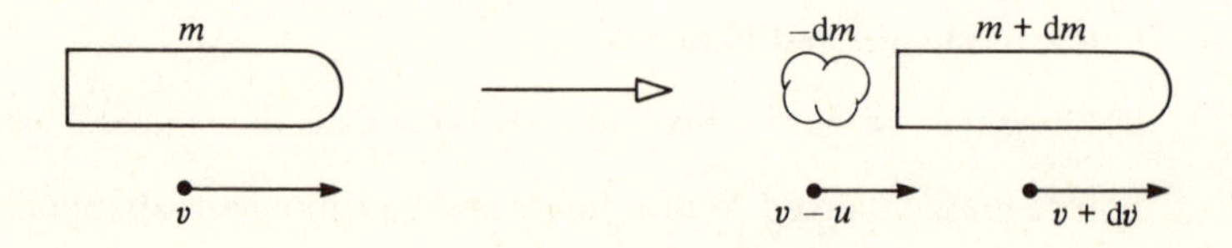

In practice dm is negative.
(i) Use conservation of momentum to establish that, in general, $\ln(m_i/m_f) = u^{-1}(v_f - v_i)$.

B11 (i) The velocity of m after the impact $= 2M(2gh)^{1/2}/(m+M)$ and the fraction of energy transferred $= 4Mm/(m+M)^2$.
(ii) Show that the maximum energy is transferred to the third ball when m is chosen so as to maximize $m(m+M)^{-1}(m+\mu)^{-1}$.

B12 Use two-dimensional vectors, taking the wind direction along $\theta = 0$, to establish that $\dot{\theta}$ satisfies $r^2\dot{\theta}^2 + (2kVr\sin\theta)\dot{\theta} + (k^2-1)V^2 = 0$, and then evaluate

$$\int_0^{2\pi} (\dot{\theta})^{-1}\, \mathrm{d}\theta$$

to order k^2.

B13 Decompose the motion into that of the centre of mass and that in the centre of mass system. The diagram shows the situation in the centre of mass system (*a*) just before, and (*b*) just after, the rope tightens.

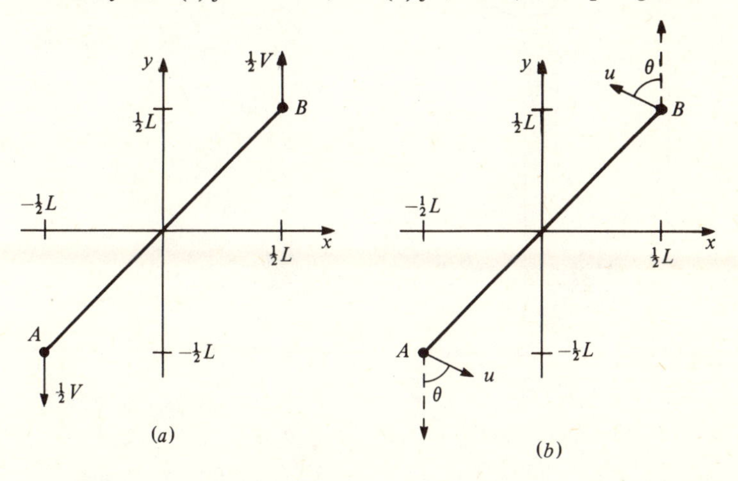

Show that $u = \frac{1}{2}V$ and that $\theta = \frac{1}{2}\pi$.

C Circular and rotational motion

C1 $mr\omega^2 = k(r-l)$, where k is the spring-constant and l the spring's unstretched length.

C2 Consider the pressure difference needed across a length dr of the water column to keep the water it contains in circular motion.

C3 $T=(l-l_0)mg/l_1; T=ml\omega^2$, where T is the tension in the string.

C4 (i) $m(l\theta)^2\omega=J=m(l_0\theta_0)^2\omega_0$.
(ii) When the string is shortened by $-\mathrm{d}l$, show that $\mathrm{d}W=(-T\sin\theta)\times(\frac{1}{2}\mathrm{d}l\sin\theta)$, and hence that

$$W=-\int_{l_0}^{l}\frac{J\omega}{2l}\,\mathrm{d}l.$$

C5

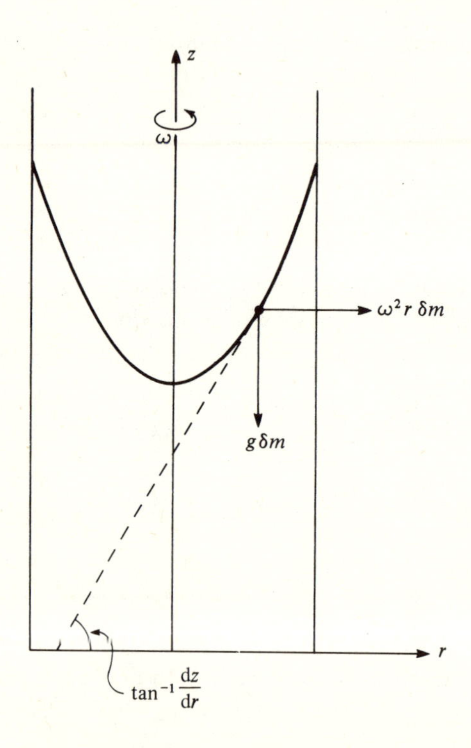

(i) The net force on an element of the free surface must be perpendicular to the tangent to the surface, i.e.

$$g\,\delta m\,\frac{\mathrm{d}z}{\mathrm{d}r}=\delta m\,\omega^2 r,$$

which yields $c=\omega^2/2g$.
(ii) Air inside the paraboloid which is also in the vessel must have volume $\frac{1}{3}\pi a^2 h$.

C6 Use $\frac{1}{2}(\frac{1}{3}Ml^2)\omega^2=Mg(\frac{1}{2}l)(1-\cos\theta)$.

C7 Show that the ratio of angular velocities (in space) of men and table is $\frac{1}{2}$.

C8

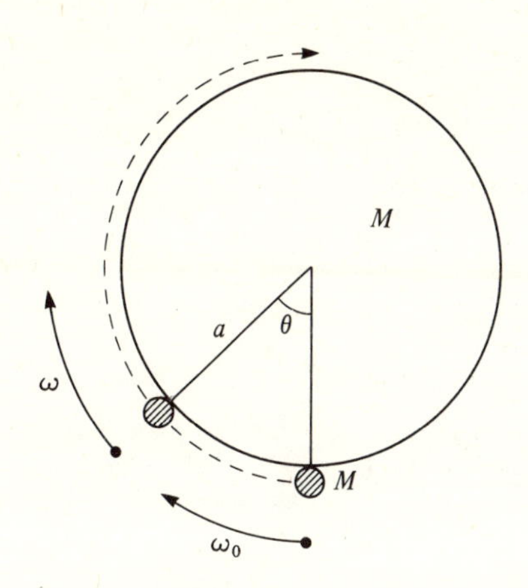

(i) Use conservation of energy to show that $2\omega^2 + \omega_0^2(1 - \cos\theta) = 2\omega_0^2$.

Use $T = \int_0^{\frac{1}{2}\pi} (\omega)^{-1}\, d\theta$.

C9

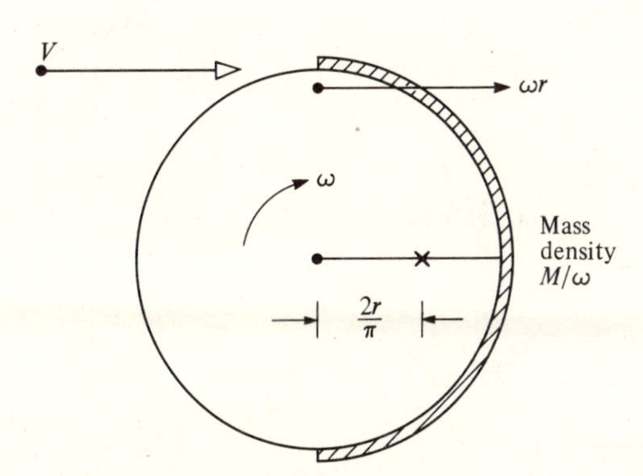

(i) Show that there is a gravitational torque of $2Mgr/\omega$, and an impulsive one of $Mr(V - \omega r)$. Power $= Mr(V\omega - \omega^2 r + 2g)$.

C10 (i) $I = 2\pi(\frac{4}{3})(\frac{1}{4}ka^4)$, $m = 2\pi(2)(\frac{1}{2}ka^2)$.
(ii) Show $\Delta T = (2\pi/I\omega^3)\Delta E$ for each year.

C11 (*a*) Consider both linear and rotational impulses.

(*b*) Show $P + R : Py = \int x\rho\, dx : \int x^2\rho\, dx = \frac{5}{6}l^2 : \frac{7}{12}l^3$.

C12 (i) Show that the radial acceleration $\propto r^{-1}$ and hence that $\dot{r}^2 = R^2\omega_0^2(1 - R^2 r^{-2})$.

C13

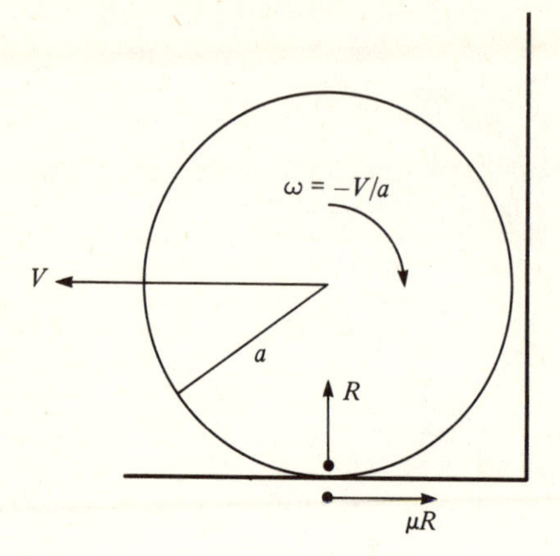

After the rebound, consider linear and rotational accelerations and show

$$\frac{M}{I}\frac{\mathrm{d}v}{\mathrm{d}\omega} = -\frac{1}{a},$$

where v and ω are the linear and rotational velocities at a general time, with $v = V = (4gh/3)^{1/2}$ and $\omega = -V/a$ just after the collision (see figure). Skidding stops when $u = v = a\omega$. The moment of inertia of the roller I is $\frac{1}{2}Ma^2$.

D Gravitational and circular orbits

D1 Show $gR_{\mathrm{E}}^2 = r^3\omega^2$.

D2 Show $\frac{1}{2}d^3(52\omega_E)^2 = GM_{\mathrm{S}} = R_{\mathrm{ES}}^3\omega_{\mathrm{E}}^2$.

D3

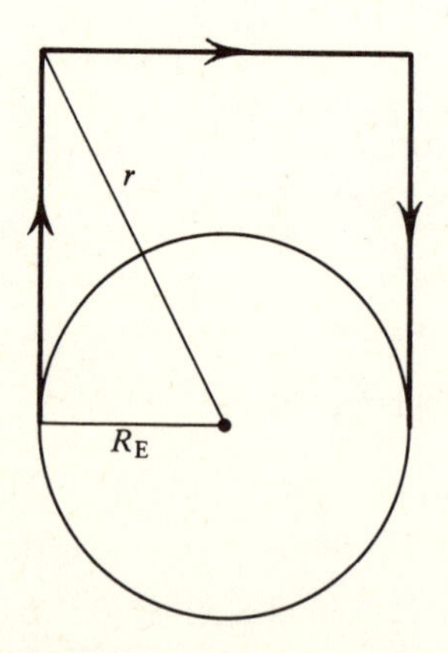

Radius r of satellite's orbit $= 4.2 \times 10^7\,\mathrm{m}$.

D4 For all r,

$$\frac{e^2}{4\pi\epsilon_0 r^2} = \frac{Gm^2}{r^2} \ .$$

D5 (i) Show

$$\frac{1}{v}\frac{\mathrm{d}v}{\mathrm{d}t} = \frac{-r}{gR_{\mathrm{E}}^2 m}\frac{\mathrm{d}E}{\mathrm{d}t},$$

with $r \approx R_{\mathrm{E}}$.

D6 Establish that $GM = \omega^2 r^3, J = mr^2\omega, E = -GMm/2r$.

D7 Show K.E. = −P.E. for an orbit of any radius.

D8 Escape: $\frac{1}{2}m(V_{\mathrm{E}} + V_{\mathrm{A}})^2 = GMm/R_{\mathrm{ES}} = mV_{\mathrm{E}}^2$. Fall into Sun's centre: $V_{\mathrm{E}} + V_{\mathrm{B}} = 0$. Show

$$|V_{\mathrm{B}}| - V_{\mathrm{A}} = \frac{2|V_{\mathrm{B}}|V_{\mathrm{A}}}{|V_{\mathrm{B}}| + V_{\mathrm{A}}} > 0.$$

D9 Establish that

$$\ddot{x} = -\frac{Gm_1m_2}{x^2}\left(\frac{1}{m_1} + \frac{1}{m_2}\right),$$

where x is the separation of centres.

D10 (iv) Are the star's surfaces equipotentials? Which point on B has the highest potential, remembering that the potentials are negative? Establish that the required condition is

$$\tfrac{1}{2}v^2 = -\frac{2GM}{3r} - \left(-\frac{8GM}{7r}\right)$$

D11 (i) $\dfrac{\Delta\lambda}{\lambda} = \dfrac{GM}{R_{\mathrm{S}}c^2}$.

D12 (i) $T = \dfrac{2\pi r^{3/2}}{(GM)^{1/2}}\left(1 - \dfrac{3GM}{rc^2} + \ldots\right)$.

(ii) Show that Mercury makes 416 orbits per century.

D13

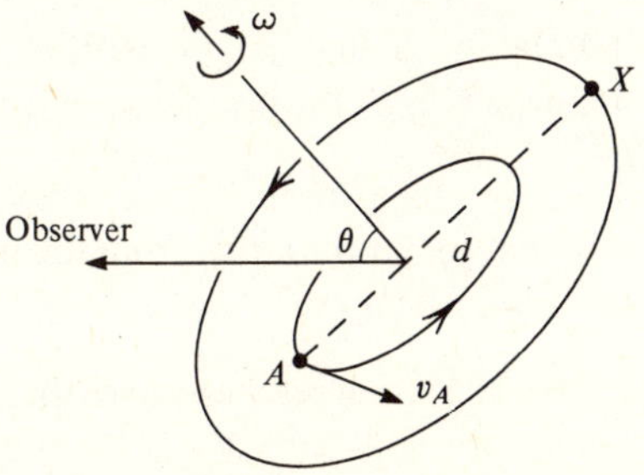

(iii) $P(X = \text{black hole}) = P(\sin^3\theta < 0.512)$.

E Simple harmonic motion

E1 (*b*) (i) Total energy $= \frac{1}{2}m\omega^2a^2$.

E2 (ii) $(A^2-16)^{1/2}\sin(1) = 4\cos(1)+3$.

E3 $2fl$ of energy lost in a time $2\pi(m/k)^{1/2}$.

E4 (i) Contact is lost when the platform reaction is zero.
(ii) Maximum reaction $=2mg$.
(iii) Contact lost when $\sin(2\omega_1 t)=0.25$. Speed is then $0.2\omega_1\cos(2\omega_1 t)$.

E5 (*b*) $m\ddot{x} = -V\rho_0 g[(x/D)-1]$. Set $x-D=y$.

E6 Compare the potential shapes with $V(x)=\frac{1}{2}kx^2$.

E7

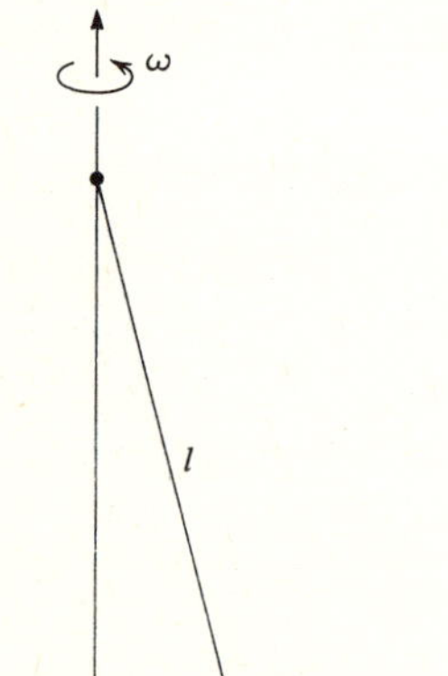

(i) $\left(\frac{T'}{T}\right)^2 = \frac{(g/l)}{(g/l)-\omega^2}$.

E8 Introduce symbols for the number of threads (n), the tension in each thread (T), the radius (a) and mass (M) of the disc; show that the restoring couple is $Ta^2n\theta/l$.

E9 Obtain two expressions for the tension in the string from the motions of the mass and flywheel. $I=2.34\times10^{-3}\,\text{kg m}^2$.

E10 Kinetic energy is conserved when the string reaches/leaves the lower peg.

E11 (*c*) Show $m\ddot{x} = k(z - x) + b(\dot{z} - \dot{x})$ and hence that $B\omega_0 \cos \phi = A\omega$ and $B\omega_0 \sin \phi = -Ak/b$.
(*d*) At high ω, equation of motion is $B\omega^2 m \cos(\omega t + \phi) \approx Ab\omega \sin(\omega t)$.

E12 (i) Momentum is conserved when the second mass is added.

E13

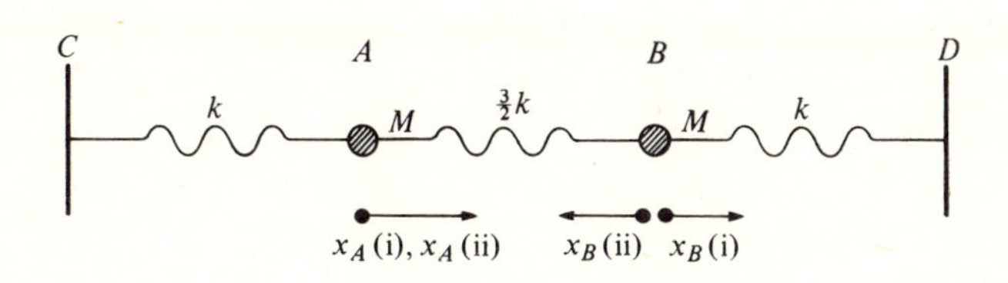

(*b*) Superpose the results from (*a*), both physically and mathematically to show that, with $\omega_0^2 = k/M, x_A = \frac{1}{2}x_0 (\cos \omega_0 t + \cos 2\omega_0 t)$.
(*c*) Establish the initial direction in which B moves.

F Waves

F1 The cork has vertical motion only with $f = 0.395$ Hz. The surface tension effect is negligible.

F2 (*a*) The linear density is $2/1.2\,\text{kg m}^{-1}$ in each case. Case (ii) is analogous to sound waves in a solid.
(*b*) Consider the dependences of the length, the tension and the linear density on λ.

F3 (*b*) Tension $= 2.2 \times 10^9 A$ newtons, where A is the string's cross-sectional area.

F4 (*a*) Consider whether, in view of the effect of the Plasticine, the fork frequency is greater than or less than 250 Hz.
(*b*) $\lambda = 0.6$ m. Air is essentially diatomic, helium monatomic. Show this implies $\gamma(\text{Air})/\gamma(\text{Helium}) = 21/25$.

F5 (i) Show that at certain times both waves pass through zero together at the midway point, as well as at the sources.
(ii) Show advance/retardation needed at the midway point is $\frac{1}{8}$ ms.

F6 If t_n is the time of the nth echo, establish that $[f(t_n)]^{-1} = \mathrm{d}t_n/\mathrm{d}n$.

F7 Show that the phase difference $\phi = \frac{2}{3}\pi[2(h^2 + 6.25)^{1/2} - 5]10^5$, where $10^5 h$ is the height of the layer, and then consider $\mathrm{d}\phi/\mathrm{d}t$.

F8 (*b*) (ii) Velocity of end of solar equator $= \pm 1.97 \times 10^3 \, \mathrm{m\,s^{-1}}$.

F9 $f' = f_0 [(v+u)/(v-u)]$. Beats at frequency $2uf/(v-u)$.

F10 (i) Consider the initial and ultimate values of f. For (iii) use linear interpolation.

F11

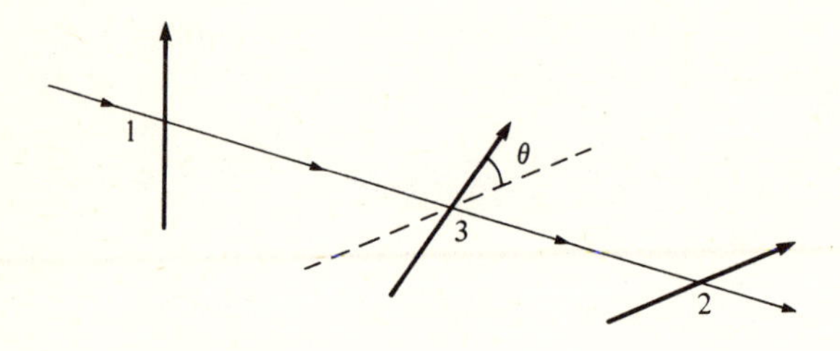

Show that the intensity incident on the last polaroid in the line is $0.32 \times 0.64 \times \sin^2\theta$.

F12

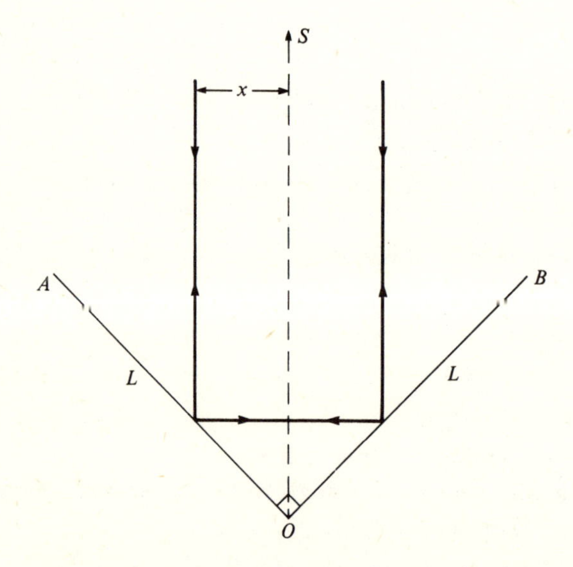

Show that the total path length inside the triangle AOB is $(2)^{1/2}L$, independent of the lateral position x of the incident sound, and independent of possible phase changes on reflection.

F13 Write $y(x,t) = A \cos(\omega t + kx + \phi)$ and $4\omega + 7k + \phi = z$, say. Then consider $\cos(z+k) + \cos(z-k)$ and similar expressions. Note that $y(x + (\omega/k)T, t - T) = y(x,t)$.

G Geometrical optics

G1 (ii) Effective object distance $150-\frac{1}{4}h$.

G2

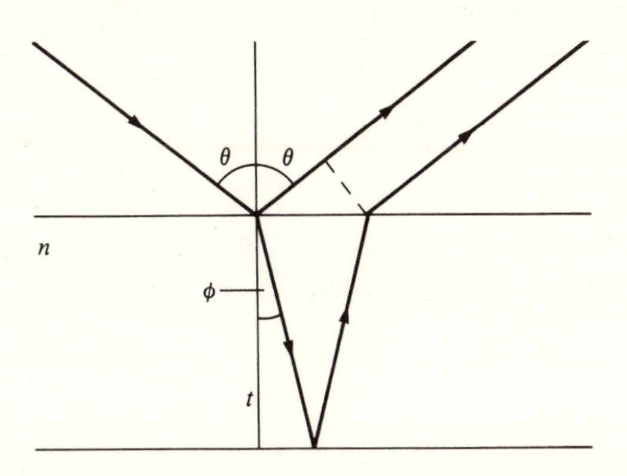

Path difference $=2nt\sec\phi-2t\tan\phi\sin\theta$, where $\sin\theta=n\sin\phi$. A phase change of π occurs on reflection from the upper surface.

G3 $2L/c=(m+\frac{1}{2})/f$; show $m=1$.

G4

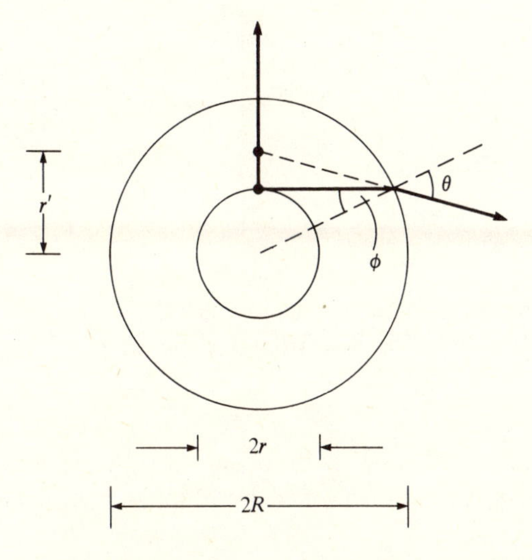

(*a*) (ii) $r'=r+(R^2-r^2)^{1/2}\tan(\theta-\phi)$ with $\sin\phi=r/R$ and $\sin\theta=n_G\sin\phi$. For $r\ll R$, $r'\approx n_G r$.
(*b*) $r'\approx(n_G/n_W)r$; $r''\approx n_W r'$.

G5 Establish that $dn=0.66\,dD$ and hence that the angular separation of the lines $=0.34°$.

G6 (i) Deviation $= \left(\theta - \dfrac{\theta}{n}\right) + \left[n\left(A - \dfrac{\theta}{n}\right) - \left(A - \dfrac{\theta}{n}\right)\right].$

G8 Intermediate image 18.52 mm from eyepiece.
(iv) $\Delta u = (u/v)^2 \Delta v$ with $\Delta v = 1.48$ mm.

G10 (iii) Show that
$$\frac{16}{76} = \frac{h'}{h} = \frac{44.4}{f'},$$
where h' is the distance from the axis at which a ray, initially parallel to the axis and distant h from it, passes through the diverging lens.

G11 (i) Show
$$\frac{n_0(1+\alpha y)}{\left[1+\left(\dfrac{\mathrm{d}y}{\mathrm{d}x}\right)^2\right]^{1/2}} = n_0,$$
and hence that $\cosh(\alpha x) = 1 + \alpha y$ and $x \approx (2y/\alpha)^{1/2}$.

G12 (*b*)

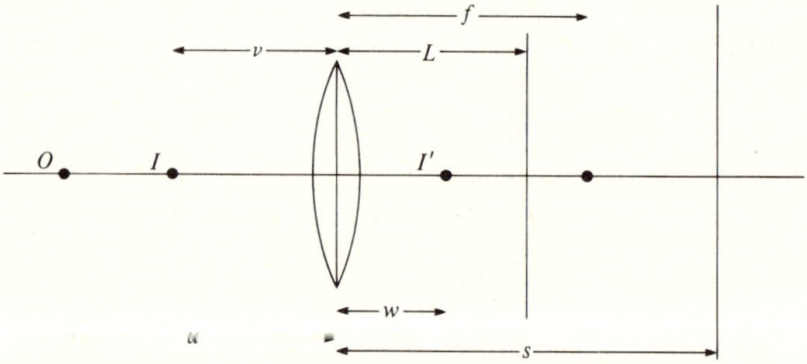

Let the image I' formed after one passage through the lens and reflection from the plane mirror be w from the lens. Show
$$w = \frac{fv}{v-f}, \quad 2L - w = \frac{fu}{u-f}.$$
Write $u = u' - s$ and $v = v' - s$ and use the requirement that u' and v' must satisfy an equation of the form $u'^{-1} + v'^{-1} = F^{-1}$ to obtain the stated results.

G13 (*a*) (i) $n_G \sin\beta = \cos\theta$ and $n_G \sin(\theta + \beta) > n_W$ lead to
$$\cos^2\theta < \frac{n_G^2 - n_W^2}{1 + n_G^2 - 2n_W}.$$
Here β is the angle, in the glass, between the ray and the normal at the point of entry into the prism.

H Interference and diffraction

H1 (*a*) (i) The additional optical path length in the film must be 15 λ.
(iii) Consider the special property possessed by the zero-order fringe and no other fringe.
(*b*) Show that $y_n = 2000\,n\lambda$ and thus that two colours overlap on the screen if $n_1\lambda_1 = n_2\lambda_2$.

H2

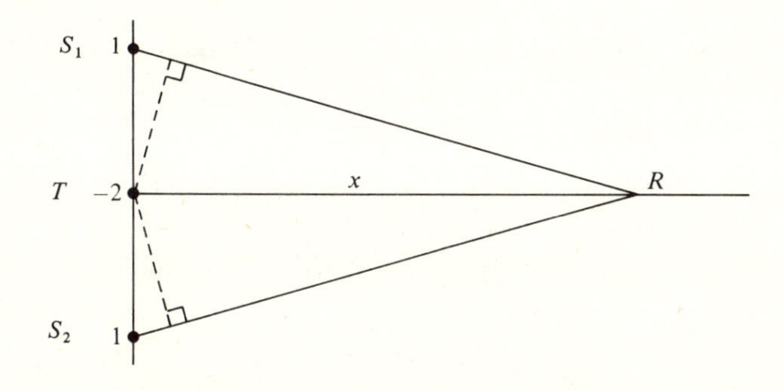

(*a*) The sources have amplitudes 1, −2, 1. Show that the path difference involved is $\approx (0.25)^2/2x$, where $x = RT$.
(*b*) Power $\propto$ (Resultant amplitude)^2.

H3

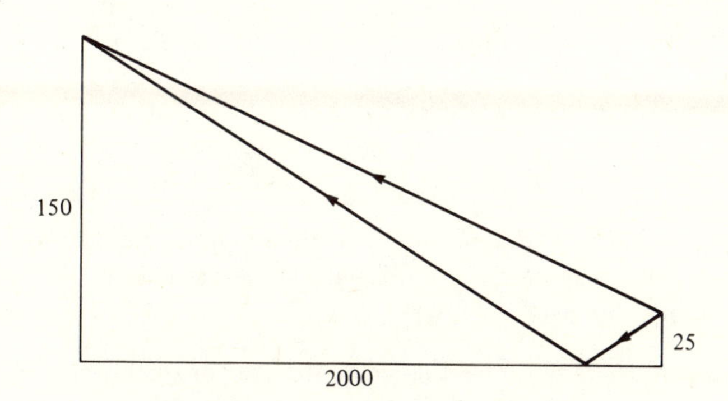

Since e.m. waves travel more slowly in water than in air, there will be a phase change of π on reflection from the sea. From geometry show that the paths have lengths $[(2000)^2 + (150+25)^2]^{1/2}$ and $[(2000)^2 + (150-25)^2]^{1/2}$, and hence that

$$n\lambda \approx \frac{1}{2}\left(\frac{300 \times 50}{2000}\right).$$

H4 (i) The plates separate by $\frac{1}{2} \times 18\lambda$ in 225 mm.
(ii) There is a phase change of π on reflection at the lower surface.

H5 It is the difference between the two coefficients of expansivity which determines the change in the air gap. The difference can be positive or negative.

H6 For the composite line given, $n\lambda = 1380 \times 10^{-9}$ m. The only physically possible cases for red and violet light are $n_R = 2, \lambda_R = 690$ nm and $n_V = 3, \lambda_v = 460$ nm. In all cases $n\lambda \leqslant 3333 \times 10^{-9}$ m.

H8 (*a*) For $\lambda \approx 600$ nm and a pupil diameter of 1 mm, $\Delta\theta \approx 6.0 \times 10^{-4}$ rad. Since 2 min of arc $= 5.8 \times 10^{-4}$ rad., the resolution is structure-limited.
(*b*) $\Delta\theta$ (eye) $\approx$ magnification $\times$ $\Delta\theta$ (objective).

H9 (*a*) (ii) With L ahead in phase by $\frac{1}{2}\pi$ the radiation from R which in (i) gave constructive interference needs to gain a further $\frac{1}{4}\lambda$ in path length, compared to that from L, in order to again produce constructive interference. (iii) L gains 2π every second.
(*b*) Orthogonal electric fields cannot interfere.

H10

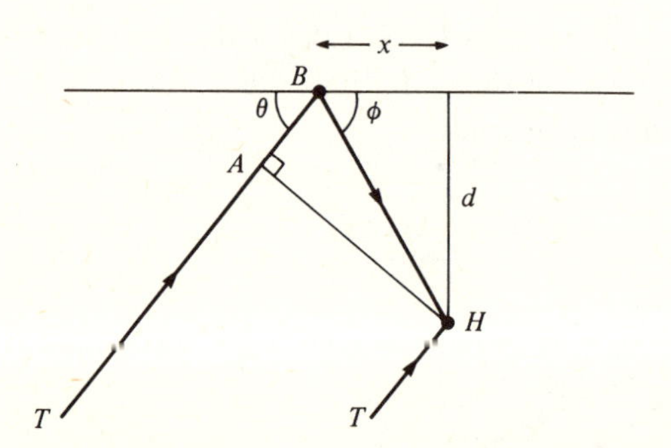

Extra path length $p = AB + BH$ and fluctuation rate $= \lambda^{-1}\, \mathrm{d}p/\mathrm{d}t$. Also, $x = d \cot\phi$ and $v = \mathrm{d}x/\mathrm{d}t$. Show $\mathrm{d}p/\mathrm{d}t = 0$ when $\phi = \theta$. Relate $\mathrm{d}p/\mathrm{d}t$ to $\mathrm{d}\phi/\mathrm{d}t$, and hence to v, when $\phi = \frac{1}{2}\pi$.

H11 Optical path difference for light reflected from upper and lower surfaces of film is $p = 2t(n \sec\theta - 2^{-1/2} \tan\theta)$ where $2^{-1/2} = n \sin\theta$ (See **G2**). $\Delta p = \lambda$ when $\Delta t = 4 \times 10^{-3}\, \alpha$.

H12 The radial blurring shows that $250\,\lambda_1 = 249\frac{1}{2}\lambda_2$. The central fringe disappears when $2d/\lambda_1 = 2d/\lambda_2 + n + \frac{1}{2}$.

H13 (*a*) A minimum of the pattern requires that the effect of the light from one half of the slit exactly cancels that due to light from the other half, i.e., the light from the centre of the slit is exactly out of phase with that from an extreme edge of the slit by the time it reaches the distant screen.

(*b*)

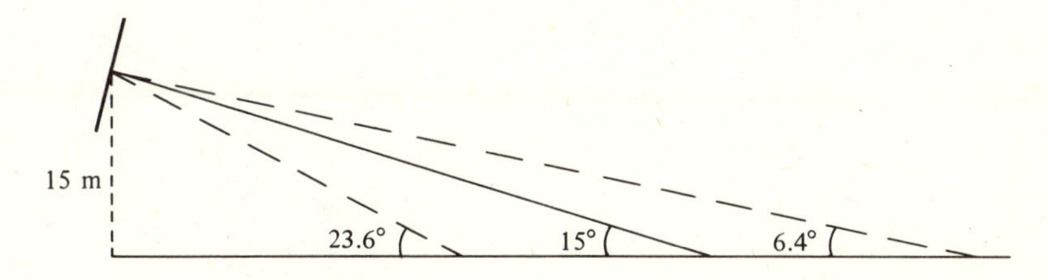

The beam spreads between $15^\circ \pm 8.6^\circ$. Length of road covered = $15(\cot 6.4^\circ - \cot 23.6^\circ)$.

I Structure and properties of solids

I1 Show that the volume of one molecule is $5.2 \times 10^{-28}\,\mathrm{m}^3$, and hence that the depth of the layer is $3.7 \times 10^{-10}\,\mathrm{m}$.

I2 $\lambda = 1.24 \times 10^{-10}\,\mathrm{m}$. Use $n\lambda = 2d \sin\theta$ with $2\theta \geqslant \theta_0$ and the maximum spacing d available in a cubic structure, namely the cube side.

I3 Show that the number of atoms in a unit cube is four and hence that the cube has side $4.08 \times 10^{-10}\,\mathrm{m}$. Show that the inter-plane spacing is $\frac{1}{3}$ of the body diagonal length i.e. is $2.36 \times 10^{-10}\,\mathrm{m}$.

I4 (i) Use $\mathrm{d}V/\mathrm{d}x = 0$ at $x = x_0$.
(ii) Let the interatomic separation be $x_0 + y$, so that the strain is y/x_0. Calculate the corresponding force to first order in y/x_0.
(iii) The chain breaks when the applied force becomes larger than the force needed to sustain the present extension, i.e. when the force is a maximum, i.e. $\mathrm{d}^2 V/\mathrm{d}x^2 = 0$.

I7 Note that for small loads, but such that σ_{mf} is exceeded, the matrix strain is equal to the carbon fibre strain, and hence $\sigma_\mathrm{c} > \sigma_\mathrm{m}$. For larger loads the fibres fracture and the load is transferred to the matrix. If x is low enough ($< x_0$) the load so transferred is small enough that σ_m does not exceed σ_{mt} and the composite does not yield until further loaded to the point at which $\bar{\sigma}$ (the average stress in the composite) equals $(1-x)\sigma_{\mathrm{mt}}$. Whilst this condition is satisfied $\bar{\sigma}$ decreases as x increases. At $x = x_0$, for a composite of cross-sectional area A, $A\sigma_{\mathrm{ct}} x_0 = A(1-x_0)(\sigma_{\mathrm{mt}} - \sigma_{\mathrm{mf}})$.

I8

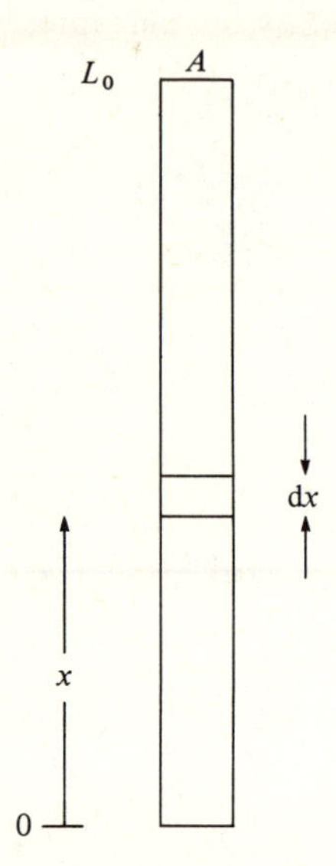

Show that a length dx, a distance x from the free end of the wire, is stretched to a length d$x(1+x\rho g/E)$.

I9

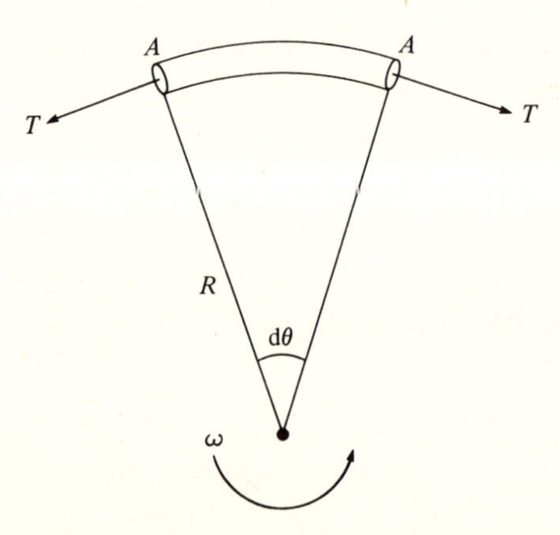

Consider an element of the ring subtending a small angle θ at its centre and establish that, if T is the tension in the ring and A its cross-sectional area, $2T(\theta/2)=R\theta A\rho R\omega^2$.

I10 If the (compressive) stress is σ, establish $x=L\alpha_1\Delta T-L\sigma/E_1$ and another similar equation.

I11 Note that the compression x of the handle gives directly the coordinate of the hammer head. Establish that $m\ddot{x} + \omega^2 x = 0$, with $\omega = (EA/ml)^{1/2}$, and hence that $x = \omega^{-1}(2gh)^{1/2} \sin \omega t$.

I12 (*a*) Stored energy $1.7 \times 10^4\ \mathrm{J\,m^{-3}}$.
(*b*) Considering unit volume $\frac{1}{2}Ee^2 = l\rho$.

I13 (i) Show that the effective 'spring-constant' k is $(10^7 \times 0.2)/(0.08 \times 0.02)\mathrm{N\,m^{-1}}$. The collision remains elastic if $x \leqslant x_\mathrm{m}$, where $x_\mathrm{m} = 0.08 \times 0.02$ and $2 \times \frac{1}{2}kx_\mathrm{m}^2 = \frac{1}{2}Mv_0^2$.

J Properties of liquids

J2 If $V(x)$ is the volume of the air when the diver is at depth x, establish the following: $V(0) = M/\rho$, where M is the mass of the cylinder; $V(x) = P_0 V(0)/(2P_0 + \rho g x)$. Establish the equation of motion for the cylinder and, by writing $\ddot{x}$ as $v(\mathrm{d}v/\mathrm{d}x)$ and integrating, obtain the stated result.

J3 (ii)

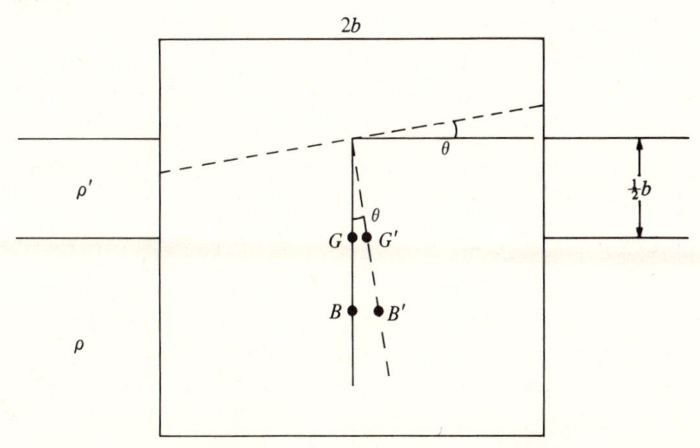

If G is the centre of gravity of the block, and B that of the displaced fluid, show that the 'capsizing couple' is $4b^2\rho' g\theta(\overline{GB})$, where $(\overline{GB})$ is the distance between G and B, equal to $15b(\rho - \rho')/16\rho$. Show that the 'restoring couple' is $2(\frac{1}{2}b^2\theta\rho' g)(\frac{2}{3}b)$.

J4 Establish that

$$\tfrac{4}{3}\pi a^3 \rho g + 2\pi(a^2 - h^2)^{1/2}\, \frac{(a^2 - h^2)^{1/2}}{a}\, \sigma = V\rho_\mathrm{L} g,$$

where V is the volume of liquid displaced and ρ_L its density.

J5

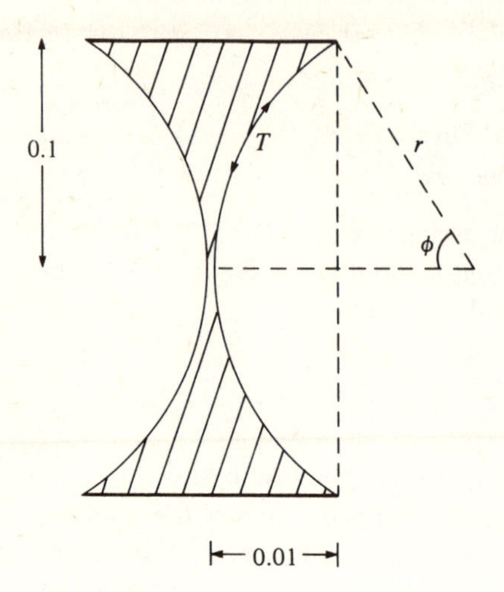

Establish that $2\gamma r = T = 15(2r\phi - 0.2)$, where $r \sin \phi = 0.1$ and $r = 0.505$ m.

J6 (i)

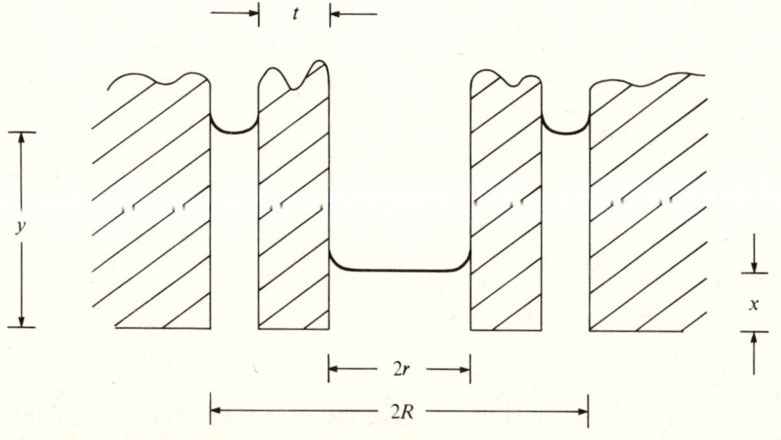

If the height difference is $y - x$, then $2\pi r\gamma = \pi r^2 \rho g x$, and $[2\pi R + 2\pi(r+t)]\gamma = \pi[R^2 - (r+t)^2]\rho g y$.

J8 Establish that
$P_i = P + 4\gamma/r_i$,
$\frac{4}{3}\pi r_i^3 P_i = n_i RT$,
$n_3 = n_1 + n_2$,
with $r_1 = a, r_2 = b, r_3 = c$.

J9 (i) Volume occupied by each molecule $= 2\sqrt{3}r^2 \times 0.1r$. Each molecule has six nearest neighbours in its own plane.
(ii) Surface energies are $\epsilon/4\sqrt{3}r^2$ and $\epsilon/3r^2$ in the two cases.

J10 Retarding couple $= \frac{1}{2}\pi\eta\omega a^4 d^{-1}$.

J11 (*b*) Show $\mathrm{d}z/\mathrm{d}V = -(\pi b^2)^{-1} - (\pi a^2)^{-1}$, and then use the expression for $\mathrm{d}V/\mathrm{d}t$ to obtain one for $\mathrm{d}z/\mathrm{d}t$.

J12 Obtain expressions for ϕ and $\mathrm{d}V/\mathrm{d}t$ in terms of the bubble radius r, and hence show that $r_0^4 - r^4 \propto t$.

J13

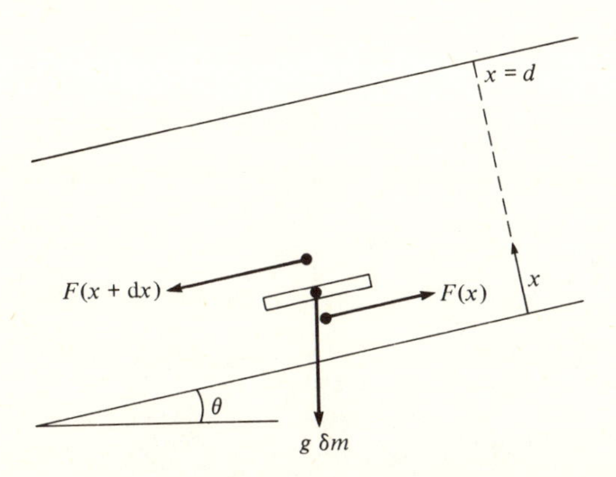

Consider the forces acting parallel to the channel bottom ($x = 0$) on a layer of oil of thickness dx, width w, and unit length. Hence establish that

$$\rho g \sin\theta + \eta \frac{\mathrm{d}^2 v}{\mathrm{d}x^2} = 0,$$

where $v(x)$ is the flow velocity at distance x from the channel bottom. Establish that $v(0) = 0$, and $\mathrm{d}v/\mathrm{d}x = 0$ at $x = d$ for physical reasons, and hence that $v(x) = \rho g \eta^{-1} \sin\theta(xd - \frac{1}{2}x^2)$.

K Properties of gases

K1 (i) $nC_p 80 - nC_v 80$.

(ii) $\int_{v_i}^{v_f} p \, \mathrm{d}v$ with $v_f/v_i = 373/293$.

K2 Establish that $\frac{3}{2}kTN = \frac{3}{2}k(T + \mathrm{d}T)(N + \mathrm{d}N) - 2kT\,\mathrm{d}N$, leading to T^3/N is constant.

K3 Establish that, for the air in the bottle,

$$\text{(cold)}\ \frac{2.5\ V_1}{278} = \frac{4\ V_2}{373}\ \text{(hot)},$$

and that $V_1 - V_2 = 3.8 \times 10^{-4}\,\text{m}^3$.

K4 Show that, if the tube has cross-sectional area A, the mass of gas it contains is

$$\frac{AN_A mp}{R} \int_0^L \frac{\mathrm{d}x}{200 + (800x/L)}.$$

K5

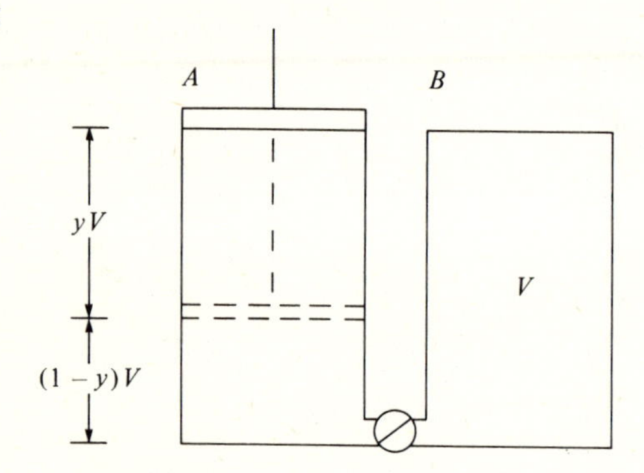

Establish the three equations;

$pV = nRT$,

$pV(2-y) = nRT_f$,

$pVy = c_{v,m}n(T_f - T)$,

where y is the fraction of cylinder A's volume through which the piston moves.

K6 Show $\rho = N_A mp/RT$ for any ideal gas and hence that

$$A = \frac{10^{-4}}{0.05 \times 10^5} \left[\left(\frac{28.8}{2}\right)\left(\frac{1}{1.05}\right) - 1\right] 9.8.$$

K7 Establish that

$$\frac{\mathrm{d}p}{p} = -\frac{N_A mg}{R}\left(\frac{\mathrm{d}x}{288 - 0.0065x}\right).$$

K8 Evaluate W given by

$$-\int_{v_i}^{v_f} (p - p_i)\,\mathrm{d}v$$

for an adiabatic compression, with $v_i = 5.18 \times 10^{-2}\,\text{m}^3$.

$W = (12.1 \times 10^3 - 4.18 \times 10^3)\,\text{J}$.

K9 (i) The greatest kinetic energy is imparted if the final pressure in the chamber plus tube is atmospheric. Evaluate

$$\int_{v_i}^{v_f} (p - p_0)\, dv \, .$$

for an adiabatic expansion.

K10 (*a*) At the equilibrium height x, both $p_{He}(x) = p_A(x) = p(x)$ and $\rho_{He}(x) = \rho_A(x)$. Use these and the gas equation to establish the equalities:

$$\frac{m_A N_A p_A(x)}{RT(x)} = \rho_A(x) = \rho_{He}(x) = \left(\frac{p_{He}(x)}{p_0}\right)^{1/\gamma} \frac{m_{He} N_A p_0}{RT(0)} \, .$$

(*b*) $p(x) = p_0(4/28.8)^{5/2}$.

K11 Show that $C_v = 21R/2$ and $pV = 5RT$. Use $-p\,dV = dU = C_v\,dT$.

K12 (ii)

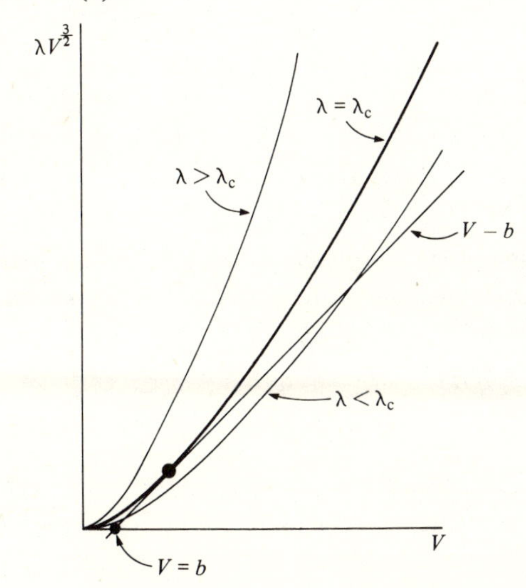

(iii) There is only one solution if the slope of $\lambda_c V^{3/2}$ [i.e. $(RT/2a)^{1/2} V^{3/2}$] at the value of V satisfying $V - b = (RT/2a)^{1/2} V^{3/2}$ has value 1. $p_c = a/27b^2$.

(iv) There do/do not exist maxima and minima on an isothermal according to whether T is less than/greater than T_c.

K13 (ii) Fractional variation is ab/V^2 in pV, which

$$\approx \frac{RT_B p^2 b^2}{V_M^3 p_0^3} \approx \left(\frac{pb}{p_0 V_M}\right)^2 .$$

L Kinetic theory and statistical physics

L1 Show that, if the subscript 1 refers to air at 300 K and 2 to helium at 20 K, then the dial records $\frac{1}{3}m_1\overline{c_1^2}n_2$ when measuring the helium pressure.

L2 (i) Show that the force is given by

$$\frac{1}{2}\sum_{1}^{1000} 2mc_i(c_i/2L).$$

(ii) $$-\frac{dN}{dt}=\frac{d}{4L}\left(\frac{N\bar{c}}{L}\right).$$

L3 $dN=-4\pi r^2(\frac{1}{4}n\bar{c})P$ with $N=\frac{4}{3}\pi r^3 n$.

L4 (i) $\phi=\frac{1}{4}n\bar{c}(C_{mol})\Delta T$.
(iii) Show, that if T_n is the temperature of the nth screen, $T_n - T_{n+1}$ is independent of n.

L5 (i) Show $$\frac{dv}{dt}=2V\left(\frac{v}{2x}\right).$$

(iii) $$\frac{3}{2}k\Delta T=-\int_{x_0}^{x} F\,dx.$$

L6 Show $I=nc_1(\frac{1}{2}mc_2^2-\frac{1}{2}mc_1^2)$ and $F=nc_1 mA(c_2-c_1)$. Subscripts 1 and 2 refer to 300 K and the temperature of the hotter side of the material respectively.

L7 (*a*) $2({}^4C_3/{}^8C_4)$.

L8 (ii) $$\Delta S=\int_0^2 \frac{20\times 25\times dt}{300+125t}.$$

L9 (i) $$k\ln\frac{W_2}{W_1}=\frac{5\times 6\times 10^3}{273}.$$

(ii) $$\Delta S=C\int_{273}^{323}\frac{dT_1}{T_1}+C\int_{373}^{323}\frac{dT_2}{T_2}.$$

L10 $$\frac{W_2}{W_1}=\frac{(3V)^{3N_A}}{(2V)^{2N_A}(V)^{N_A}}.$$

L11 (*b*) (i) $T_f=2^{-R/C_{v,m}}T$.

L12 $$\frac{E_m}{k}\left(\frac{1}{300}-\frac{1}{T}\right)=\ln 10,\ \text{where}\ E_m=\frac{2.3\times 10^6\times 18\times 10^{-3}}{6\times 10^{23}}.$$

L13 The *maximum* work is extracted when the overall change in entropy is zero. Show that this implies that the final equilibrium temperature has to be $(T_1 T_2)^{1/2}$.

M Heat transfer

M2 Establish that $(\Delta T/x)\lambda_I = l\rho_I(\mathrm{d}x/\mathrm{d}t)$. Here x is the ice thickness.

M3 $$\text{Time} \propto \frac{\text{Heat capacity}}{\text{Area} \times \text{Temperature gradient}} \propto \frac{L^3}{L^2 L^{-1}}, \text{ with } M \propto L^3.$$

M4

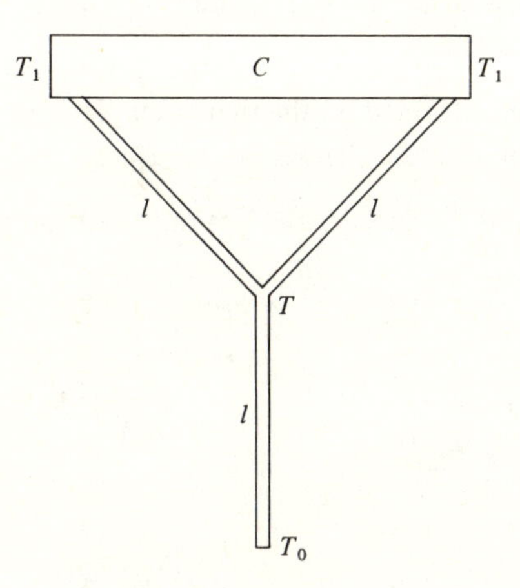

(ii) $$C\frac{\mathrm{d}T_1}{\mathrm{d}t} = 2\left(\frac{\pi d^2}{4}\right)\frac{\lambda}{l}(T - T_1).$$

M5 From the heat balance in a thin shell of thickness $\mathrm{d}r$, establish that
$$\frac{\mathrm{d}}{\mathrm{d}r}\left(-4\pi r^2\lambda\frac{\mathrm{d}T}{\mathrm{d}r}\right) = 4\pi r^2 H.$$

M6 (*a*) Show that $r(\mathrm{d}T/\mathrm{d}r)$ must be independent of r.
(*b*) (i) Justify the equation $f_\mathrm{m}cT(x) = f_\mathrm{m}cT(x + \mathrm{d}x) + \phi[T(x) - T_0]\,\mathrm{d}x$.
(ii) Show $T(x) - T_0 = (T_1 - T_0)\exp(-\phi x/f_\mathrm{m}c)$.

M7 Justify $C\,\mathrm{d}T/\mathrm{d}t = -k(T - T_0)$, with $k(T_1 - T_0) = P$.

M8 Justify $4\pi R_\mathrm{S}^2\sigma T_\mathrm{S}^4(\pi r^2/4\pi R^2) = 4\pi r^2\sigma T^4$.

M9 The angular diameters of object and image are equal. Use this to establish that
$$4\pi R_\mathrm{S}^2\sigma T_\mathrm{S}^4\left(\frac{\pi(0.05)^2}{4\pi R_\mathrm{ES}^2}\right) = 2\pi\left(\frac{0.5R_\mathrm{S}}{R_\mathrm{ES}}\right)^2\sigma T^4.$$

M10 (i) Establish that $T_{i+1}^4 - T_i^4$ is independent of i, where T_i is the absolute temperature of the ith sheet ($i = 0, 1, 2, 3$).

M12 Conduction loss = 1.50 kW, Radiation loss = 1.00 kW.

M13 Establish that $A\lambda l^{-1}(T_1 - T) = A\sigma(T^4 - T_0^4) \approx A\sigma(T - T_0)4T_0^3$.

N Electrostatics

N1 In the first change V is constant; in the second Q is constant.

N3 (v) Show that C and D must have charges -10^{-9} C and $+10^{-9}$ C respectively on their outside surfaces, and $+2 \times 10^{-9}$ C and -2×10^{-9} C respectively on their inside surfaces.

N4 (ii) The charge on the capacitor formed by the two outer plates is one-half of the difference between the charges on the plates.

N5 Final common potential = 300 V. Consider the loss of stored energy.

N6 (ii)

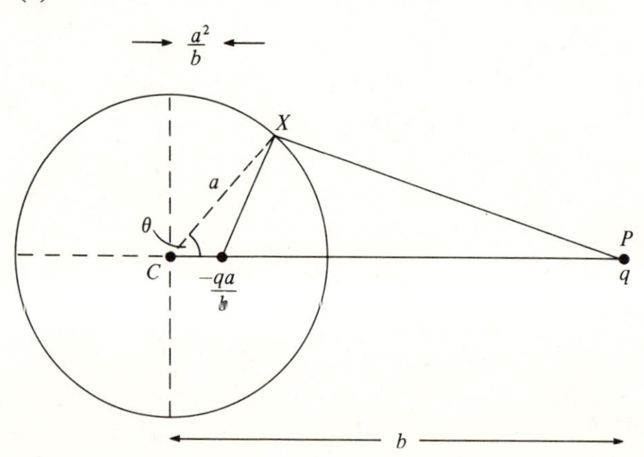

$$4\pi\epsilon_0 V_X = q[(b - a\cos\theta)^2 + (a\sin\theta)^2]^{-1/2} - qab^{-1}[(a\cos\theta - (a^2/b))^2 + (a\sin\theta)^2]^{-1/2}$$

which can be shown $\equiv 0$. The cases in (i) correspond to $\theta = \pi, 0, \frac{1}{2}\pi$.
(iii) Determine directly the forces between Q and the two charges of $\pm Qa/R$. [It is incorrect to evaluate $d\phi/dR$, since some of the R dependence of ϕ comes from the size of the image charges and not from the position of Q.]

N7 (i) Use $C = 4\pi\epsilon_0 r$, $Q = CV$ and $dQ/dt = -V/R$, to show that $dV/dt = 0$ implies $dr/dt = -(4\pi\epsilon_0 R)^{-1}$.

N8 Energy stored in both cases $=\frac{1}{2}CN^2\mathcal{E}^2$. In case (*a*) that wasted also has this value. For case (*b*):
Work done by battery $= \Delta Q\ \Sigma\ (\Delta V) = C\mathcal{E}(\mathcal{E} + 2\mathcal{E} + \ldots + N\mathcal{E})$.
Energy wasted $=\frac{1}{2}C\mathcal{E}^2\,[N(N+1) - N^2]$.

N9 (iii) Original electrostatic energy $=7.6 \times 10^{-5}$ J. Final electrostatic energy $=2^{-2/3}$ times this. The difference, 2.8×10^{-5} J, is much greater than the increase in surface energy.

N10 Voltage $\propto Q_1$ at frequency ω, and $\propto Q_1^2$ at frequency 2ω.

N11 Establish that the total stored energy is $\frac{1}{2}(2N-1)\epsilon_0 A d^{-1}(E_0 d)^2$ where $v = (2N-1)Ad$, d being the separation of successive plates.

N12 Show that, for a given value of a, the maximum potential difference that can be applied is $[a - (a^2/b)]E_0$. For (i) this has to be maximized with respect to a. For (ii) $2\pi\epsilon_0 E_0^2 a^3\,[1 - (a/b)]$ has to be maximized.

N13 (*a*) (ii) $\mathrm{d}W = -(Q^2/8\pi\epsilon_0 a^2)\,\mathrm{d}a$ must be equal to $-4\pi a^2 p\,\mathrm{d}a$.
(*b*) Show that $T = a\sigma^2/4\epsilon_0$ with $\sigma = \epsilon_0 E_0$.
(*c*) Consider the two cases of (I) ionization, and (II) tensile strength, being the limiting factor. For (I), show $W \propto a^3$, and for (II) $W \propto a^2$. Cost $\propto a^2$ in both cases.

O Direct currents

O1 Note that the central plate gains Ag^+ and loses Cu^{++} ions.

O2 (*b*) Note that the speed of the electron just before the collision is $2v_d$. Equate the heat generated to $\rho l A^{-1} I^2$.

O3

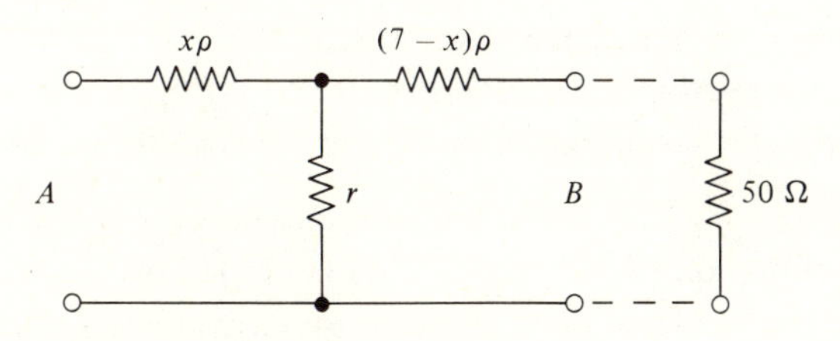

(i) Establish the three equations:
$\rho x + r = 64$,
$\rho(7 - x) + r = 70$,
$15(r + \rho x) = 16r$.
(ii) Show that the current drawn is $V/34$, and that half of it passes through the 50 Ω resistor.

O4 (ii) Consider a graphical solution, plotting, all on the same I-V_0 plot, $I_1, I_2, \frac{1}{2}(I_1+I_2)$ and $\frac{1}{2}V_0$, where V_0 is the voltage across the output terminals.

O5 Equate heat generated to radiative losses at the melting point of copper.

O8 (*a*) (ii) Note that, if the current drawn is I $(=I(r))$, then it can be delivered for a time W/EI where W is the capacity of the battery.
(*b*) Shorting is beneficial when $12/(0.11+R')=10/0.11$.

O9 Establish that $aI_A^2=bI_B^2$ and that $aI_A^2+c(I_A+I_B)^2=V$.

O10

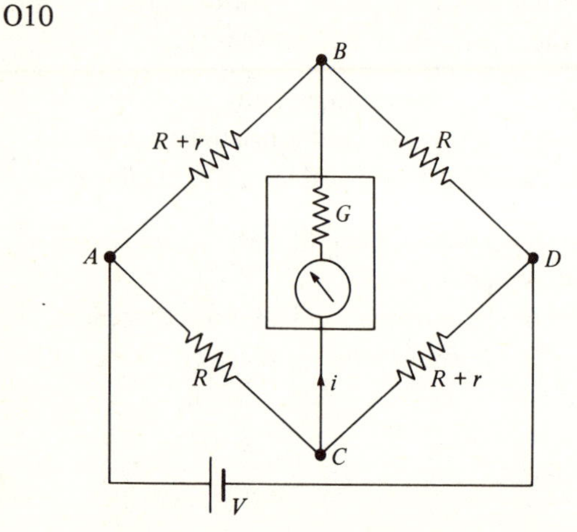

Analyse the circuit using Kirchhoff's laws, showing that the currents in branches AB and AC are $(V-Ri)/(2R+r)$ and $[V+(R+r)i]/(2R+r)$, where i is the current through the galvanometer from C to D.

O11 (*b*) Note that the current in the 8 Ω resistor is irrelevant. Express all other currents in terms of three unknowns.
(*c*) Consider the relationship of the circuit to that of a Wheatstone bridge, and find the condition for no current through the 5 Ω resistor.

O12 Show that $d+e+f=13$ and that the resistance between any pair of terminals is of the form $R=n_1+n_2+n_3(13-n_3)/13$, with all n_i integral. Use the fractional parts of the measured resistances to produce a short list of possible values of $n_3(13-n_3)$.

P Non-steady currents

P1 Consider the moment of peak current when the electrons are travelling with velocity ωa.

P2 Show that the voltmeter scales the detected signal by $\pi/2^{3/2}$ before displaying it.

P3 (*a*) (ii) Show that the circuit is equivalent to inductance $\frac{1}{2}L$ in series with capacitance $2C$. (iii) Note that a voltage $\frac{1}{2}V_0 \cos \omega t$ is applied across each component, and that the total current flowing is the algebraic sum of an inductor current and a capacitor current.

P5 Show, both electrically and graphically, that $dV_2/dt = (V_1 - V_2)/RC$.

P6 (iii) Note that period of capacitor discharge is nearly equal to 0.02 s, and that the discharge occurs with time constant 0.5 s.
(iv) Show that δ, the period each cycle for which the diode conducts, is given by $\delta = f^{-1} - T_1$, where T_1 satisfies $\exp(-T_1/RC) = \cos(2\pi f T_1)$. Then solve the equation for δ assuming $\delta \ll f \ll RC$.

P7 Either solve graphically, or let the currents be $I_1 = I_0 \cos(\omega t)$, $I_2 = I_0 \cos(\omega t + \alpha)$ and $I_3 = I_0 \cos(\omega t + \beta)$, with $I_1 = I_2 + I_3$.

P8 Show $dV/dt = -V^2/20$.

P9

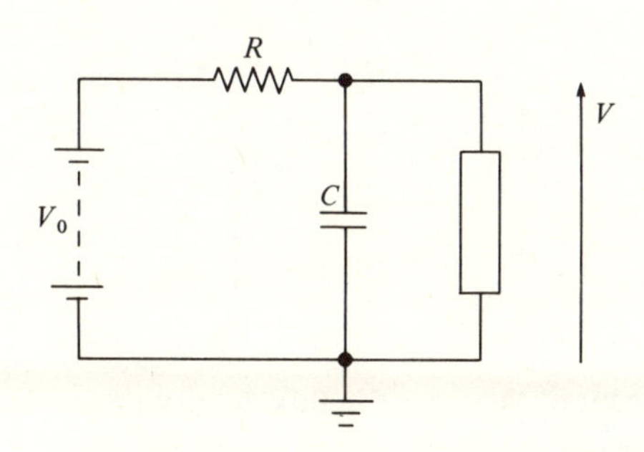

(i) Show $V_1 = V_0[1 - \exp(-T/RC)]$.
(ii) Linearity requires $t^2/2R^2C^2 < 0.01\, t/RC$ for $0 \leqslant t \leqslant T$.

P10 (*a*) Closing K momentarily sets $V_3 = +10\ V$ and it remains at this value. Note that the p.d. across C rises to $10\ V$ with time constant $C(R_3 + R_4)$.
(*b*) Show that at a time T, given by $(5/7)\exp(-T/0.07) = 1/6$, V_1 becomes $< V_2$ and V_3 becomes $-10\ V$.

P11 (*a*) Combine sets of parallel or series resistors (or capacitors).
(*b*) Obtain two first-order equations involving V_1 and V_2, and then eliminate V_2 by differentiation of one of them and substitution.

P12 Treating 0 V as '0' and 6 V as '1', establish that the unit gives output $\bar{x}$ when only one input x is used, and that it gives $\bar{x}\bar{y}$ when two inputs x and y are used.

P13 It is only necessary to establish that a proposed arrangement will sustain an 'oscillation' once it has started. Initiation of the oscillation will, in practice, occur through noise or the switching-on process.

Q Magnetic fields and currents

Q1 Show that the number of turns per unit length $\propto d^{-2}$ and that the resistance of the winding per unit length $\propto d^{-4}$.

Q2 Show that the situation with (say) X normal and Y superconducting is a stable one, provided $\frac{1}{12} < B_0/\mu_0 n V_0 < \frac{1}{2}$, but that one with both superconducting or both normal is not.

Q3

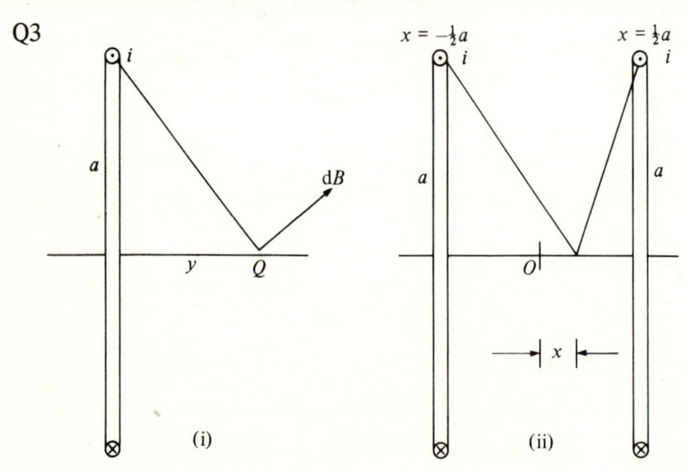

(i) Note that the component of the magnetic flux density perpendicular to the axis averages to zero.

Q4 Use the formula derived in question **Q3**.

Q6 (ii) Show $v = 0.75$ c.

Q7

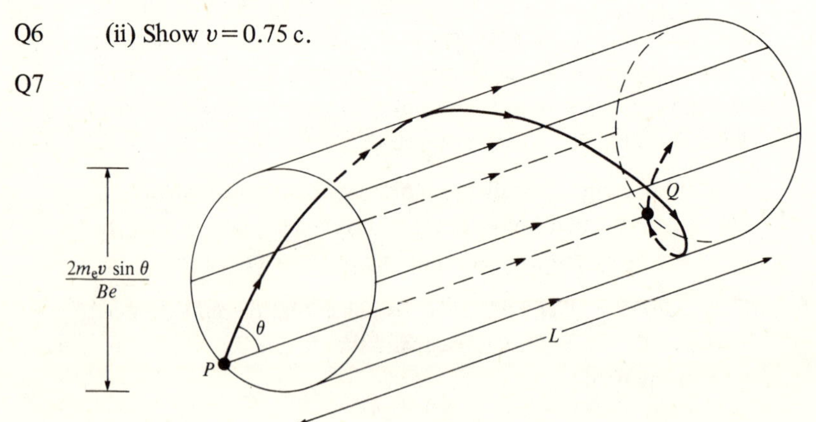

Show that whatever the radius of the helical path, $(m_e v \sin\theta)/Be$, the time to make one complete turn is $2\pi m_e/Be$, independent of θ.

Q8

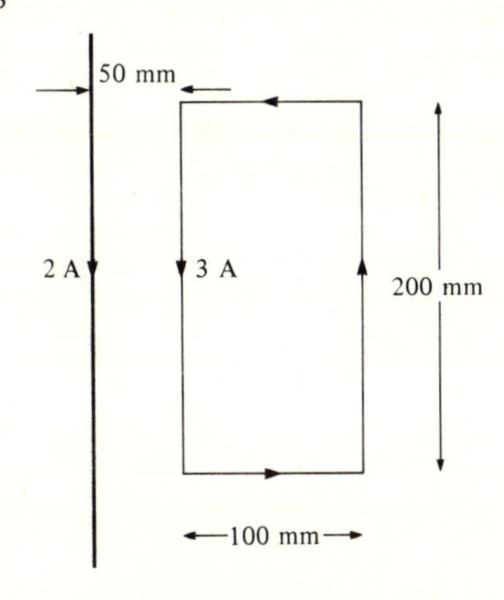

(ii) It is necessary to integrate, since the flux density varies with position along the side.

Q9 Show that both the electrostatic and the magnetic force vary inversely as the separation of the wires, and that the required condition is $\mu_0 I^2 = \epsilon_0^{-1} C^2 V^2$, where C is the capacitance per unit length.

Q10 All three cases come from consideration of the existence or otherwise of solutions of the equation $k_0 \sin\theta = k\theta$.

Q12 (ii) Treat the rings as effectively long parallel wires.
(iii) Set $h = h_0 + x$ and expand for small x/h_0.

Q13

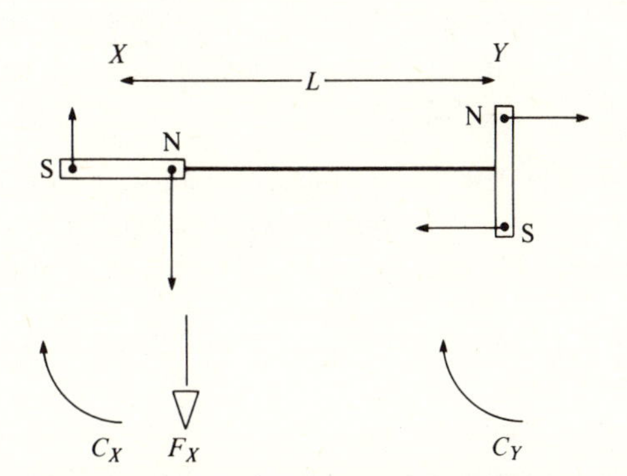

(ii) Note that X experiences a net force when in the field due to Y. Consider the effect of this and any reaction on the wooden rod.

R Electromagnetism

R2 The salt water will be a conductor moving across the Earth's vertical magnetic field.

R4 The two parts of the rod have opposing effects.

R5 (ii) Note that an annulus of radius x, width $\mathrm{d}x$ and thickness t has resistance $\rho \mathrm{d}x/2\pi xt$ between its inner and outer curved surfaces.

R6

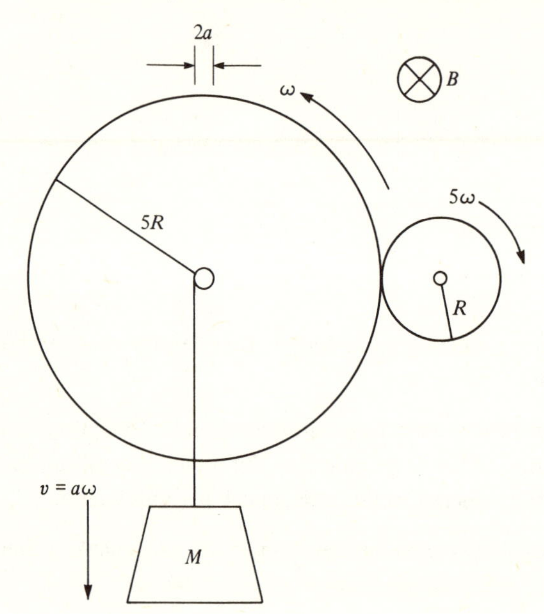

Note that the e.m.f.'s generated in the two wheels produce electric fields in the same direction at the point of contact, giving a total p.d. of $7.49 \times 10^{-4}\,\omega$ volts, where ω is the angular velocity of the larger wheel. Under terminal conditions, equate the rate of loss of potential energy to the electrical power dissipated.

R7 Use the expression for the field inside a long solenoid to show that its inductance L is given by $\mu_0 \pi a^2 n^2 l$, where l is its length. If R is the resistance of the solenoid, the time constant of the closed circuit is L/R.

R8 (ii) Equate the work done against the torque to the electrical energy dissipated in any given time.
(iii) Consider not only the resistance of the coil but also its self-inductance, and draw the resistive and inductive p.d.'s and the induced e.m.f. in a phasor diagram.

R9 (i) Establish that $-\mathrm{d}\Phi/\mathrm{d}t = L\,\mathrm{d}I/\mathrm{d}t + RI$, substitute the given forms for I and Φ, and equate separately terms which vary as $\cos \omega t$ and as $\sin \omega t$. (ii) Obtain an expression for I_0, and hence for $\frac{1}{2}RI_0^2$, which does not contain ϵ.

R10 Establish that $mv = eBr$ and that the accelerating electric field strength $E = -\frac{1}{2}r\dot{\bar{B}}$, where $\bar{B}$ is the average flux density within the orbit.

R11 (i) and (ii) Let the current through the voltmeter be i and that through the major arc of the ring be I. Then, in each case, for two different closed circuits, equate '$\Sigma r_j i_j$' to the rate of change of flux through those circuits. Use a consistent convention for current directions and the sense of circuit traversal. The voltmeter reading is given by Ri.

R12

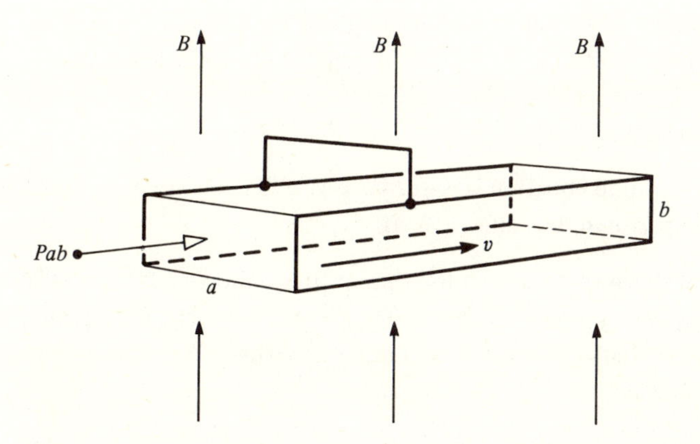

Show that the current flowing is $vBlb/\rho$ and that the 'back pressure' is vB^2l/ρ.

R13

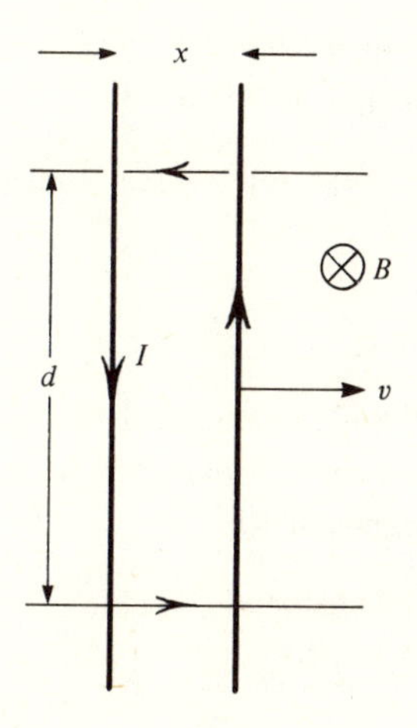

Establish that the current I induced in the two wires causes a repulsion between them of magnitude $\mu_0 I^2 d/2\pi x$, and a force on the fixed wire of BId in the same direction as v.

S Nuclei

S1 (*a*) Equate kinetic energy to electrostatic potential energy.
(*b*) The mass of a deuteron is irrelevant. Show that the radius of a gold nucleus is 6.9×10^{-15} m, and determine the energy needed for the deuteron to approach to this distance.

S2 Recall that the decay constant λ is given by $\tau_{\frac{1}{2}}^{-1} \ln 2$.

S3 Note that the reading after 2 days needs to be corrected (downwards by about 220 counts per min.) to allow for decays of type B, before the best fit line on the log-linear graph is drawn to determine λ_A. Other corrections are negligible if the readings from 5 to 15 days are used to determine λ_B.

S4 The required equation is $\mathrm{d}n/\mathrm{d}t = \lambda_X N_0 \exp(-\lambda_X t) - \lambda_Y n$, and the appropriate initial value $n(0) = 0$.

S5 Show, using the result of **S4**, that at time t the γ-count rate is given by $\lambda_A \lambda_B (\lambda_B - \lambda_A)^{-1} N_0 [\exp(-\lambda_A t) - \exp(-\lambda_B t)]$. If 11.5 minutes = t_1, note that $\lambda_A T \gg 1 \gg \lambda_B t_1$. The actual value of the counting rate is not needed.

S6 (i) Satisfy yourself that, in radioactive equilibrium, for each element in a series, except the first and last, the number of atoms present is virtually proportional to its half-life.

S7 (i) Note that 2 months $\gg$ 3.64 d or 56 s, but $\ll$ 1.91 y.
(ii) Note that 200 y $\gg$ all half-lives except that of $^{232}_{90}\mathrm{Th}$.

S8 (i) Consider adding together the equations representing reactions (*a*) and (*d*).
(ii) Beware!
(iii) Extend the method of (i).

S9 Show that only a fraction $A/(A+1)$ of the neutron's incident kinetic energy is available in the centre of mass frame to provide the necessary 4.036 MeV, where $A = 19$ is the mass number of $^{19}_{9}\mathrm{F}$.

S10 (iii) Establish that $4\mathrm{p} \rightarrow \alpha + 2\mathrm{e}^+ + 2\nu$ has a Q value of 24.7 MeV.

S11

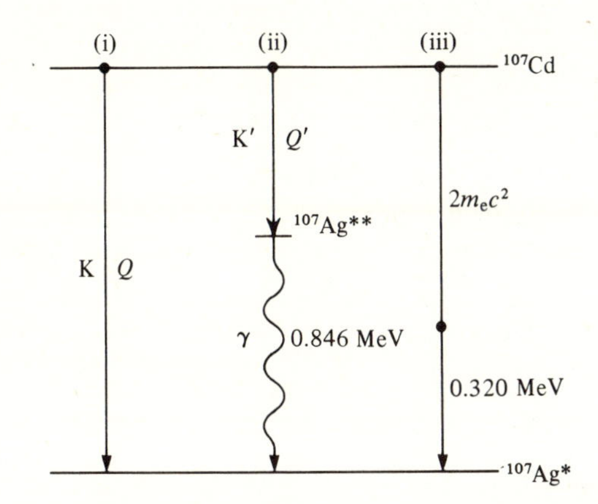

Show that the Q value of the direct capture reaction must equal $0.320\,\text{MeV} + 2m_ec^2$, that $Q' = 0.496\,\text{MeV}$, and that the neutrino momentum is very nearly equal to the Q value of the capture divided by c; hence deduce the recoil kinetic energies.

S12 (i) Denote the binding energy *per nucleon* by C, i.e. $C(A) \equiv A^{-1}B(A)$. The most stable nucleus has the maximal value for C, leading to $0.1\,A_s^{1.1} = \beta$.
(ii) Show that the necessary condition can be written as $B(A_f) < 2B(A_f/2)$, or as $C(A_f) < C(A_f/2)$, leading to $A_f^{1.1} = \beta(1 - 2^{-0.1})^{-1}$.

S13 (ii) Following the method and notation of question **S12**, show that $A_s = a_2/(2\lambda^2 a_4)$ and that $A_f = [(2^{1/3} - 1)a_2]/[(1 - 2^{-2/3})\lambda^2 a_4]$, and thus that if $A_s \approx 56, A_f \approx 79$.

T Electrons, photons and atoms

T1 Electron velocity $= 3.6 \times 10^6\,\text{m s}^{-1}$.

T2 Establish that the number n_1 of elementary charges on a drop of mass m is proportional to mV_1^{-1}, and note that before the possible quarks are introduced the various values of n_1 should be in whole number ratios (2:1:3:1:2:2). Then consider the subsequent changes in n, $\Delta n = n_2 - n_1$.

T3 Plot the given results on the basis of the equation $eV_s = hc\lambda^{-1} - \phi_0$.

T4 Show that $\lambda = h/mc$ (the so-called Compton wavelength of the particle).

T5 At a distance r from the lamp, the photons produced in time Δt occupy a volume $4\pi r^2 c\Delta t$.

T6 Establish that $T_n = (14.56 - 13.58\,n^{-2})\,\text{eV}$, where $n(=1, 2, \ldots)$ corresponds to the various energy levels in hydrogen.

T7 Establish the equations $Ze^2 = 4\pi\epsilon_0 rmv^2$ and $h = 2\pi mvr$, and that, at the limiting value of Z, $r = R = 1.7 \times 10^{-15} Z^{1/3}\,\text{m}$.

T8

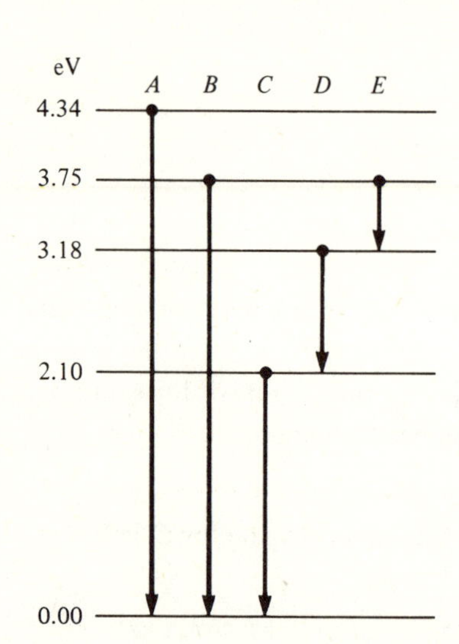

Show that each line corresponds to an energy equal to the difference between two of the levels shown.

T9 (i) Defining $y_A = \Delta E_A / n_A$, where ΔE is the energy change corresponding to any particular line and n the order of the spectrum, note that $y_A = y_B + y_E$ and that $2y_C = y_A + y_D$.
(ii) Establish that there is no difference in level energies which would correspond to $n_F = 1$ and a line in this angular range. Note, however, that $4.86 > (9.95 - 5.11) > 4.68$.

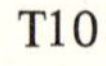
T10

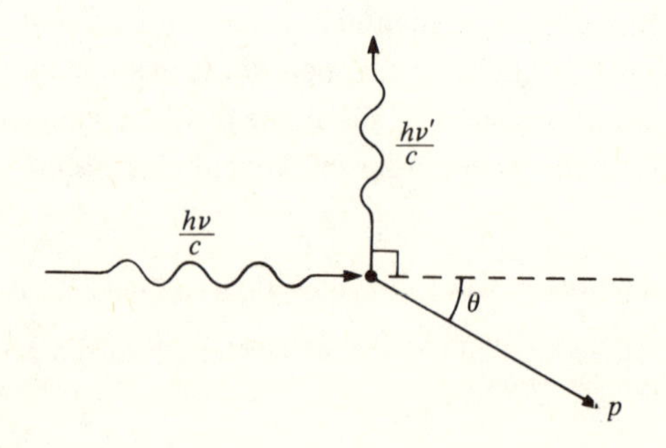

Use total energy conservation (including the rest mass of the initial electron) and momentum conservation along and perpendicular to the initial photon direction. Note that $E = pc$ for photons.

T11 (i) Show that the wave function ψ has the form $A \sin kx$, with the allowed values of k restricted by $ka = n\pi$, $n = 1, 2, \ldots$
(iii) Recall that

$$P = \left(\int_0^\epsilon |\psi|^2 \, dx \right) \Big/ \left(\int_0^a |\psi|^2 \, dx \right).$$

T12 Establish that $\lambda_1 = 2a/n$ and $\lambda_2 = 2b/m$ where m and n are positive integers, and then substitute ψ into the two-dimensional Schrödinger equation with $m = n = 1$.

T13 (i) Equate separately, terms depending upon x^2, and terms independent of x, on the two sides of the Schrödinger equation. $\alpha = \pi(km)^{1/2}/h$.

U Shorts

U3 Consider the connection with critically refracted light.

U4 Apply momentum conservation.

U5 Show that $P_1 \propto A^{-1}$ and $P_2 \propto A$, where A is the cross-sectional area of the cables, and find the condition which minimizes $P_1 + P_2$.

U6 Note that it is the frictional couple that limits the rotational speed when the motor is working.

U8 Section 2 moves down with velocity Va/b.

U12 Since the drop is free-falling, the period cannot depend upon g.

U13 Consider the connection with question **M3**.

V Longs

V1 Consider the sunlight falling on the Moon, and find the fraction of it reaching unit area of the Earth after being diffusely reflected. Show that the ratio is $\alpha R_{\mathrm{M}}^2/2R_{\mathrm{EM}}^2$ where α ($=0.1$ say) is the fraction reflected. Note that the Moon's diameter subtends 9×10^{-3} rad at the Earth's surface.

V2 For each sub-pile, consider the position of the centre of gravity in relation to its support.

V3 (ii) Note that changing the on/off status of the lamp changes the oddness/evenness of the number of switches which are up.

V4

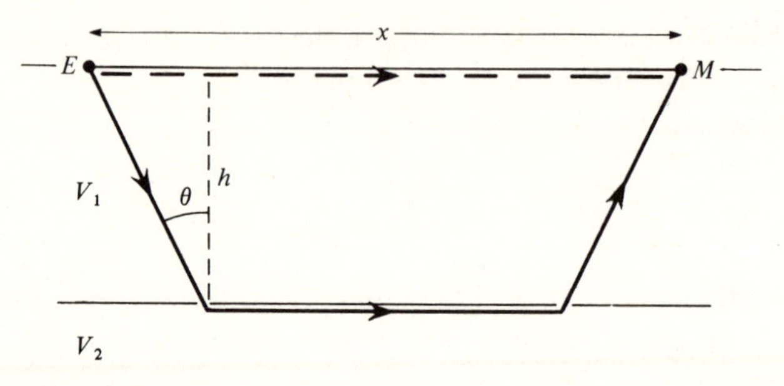

To be refracted along the interface the wave must be incident upon the interface at $\theta = \sin^{-1}(V_1/V_2)$ to the common normal. A corresponding condition applies when the wave leaves the lower rock.
$V_1^{-1}x = V_1^{-1}2h \sec\theta + V_2^{-1}(x - 2h\tan\theta)$.

V5 The current flowing before the source is disconnected is $0.5\,(1 - \exp(-40\,t))$. Contacts close at 12.8 ms. At 20 ms the current is 0.275 A. Contacts open at 45.3 ms.

V6 (ii) Show that when a ball has fallen through a vertical height h the speed v of its centre of mass is given by $v^2 = 2gh(1 + k^2/r^2)^{-1}$, where k is its radius of gyration.

V7 Introduce a as the length of a side of the base.

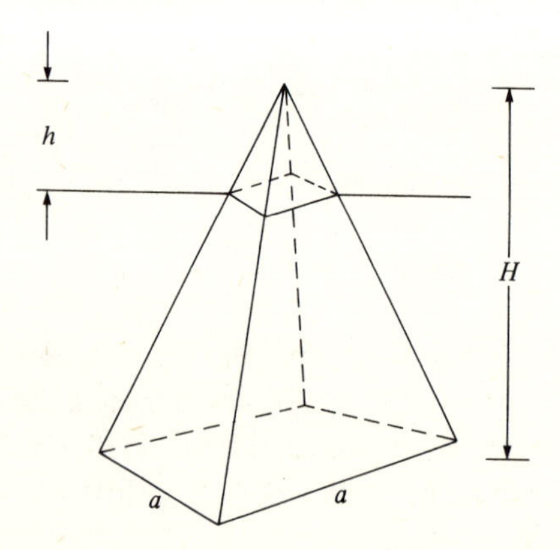

Show that $\omega^2 = 3\rho_W gh^2/\rho_I H^3$, where $h = 10$ m, H = total height of the iceberg, and $(H^3 - h^3)\rho_W = H^3\rho_I$.

V8 The total charge on the plates is unchanged during the filling and emptying.

V9 Convince yourself that it is only the East–West component of the traffic momentum that matters and that, taking the Earth's actual direction of rotation as positive, the angular momentum of the traffic about the Earth's axis will increase, and hence that of the Earth decrease.

V10 Establish that in the horizontal limb the pressure has the form $P(x)=C+\frac{1}{2}\rho\omega^2(x_0+x)^2$, but with different values of C and ρ for each liquid.

V11 Either (i) integrate both equations of motion, or (ii) integrate the transverse one and use energy conservation. The current ceases when $v_x(d)=0$, or alternatively the total energy is $p_y^2/2m_e$. The x-direction is that of the electric field; the y-direction is perpendicular to both this and the magnetic field.

V12

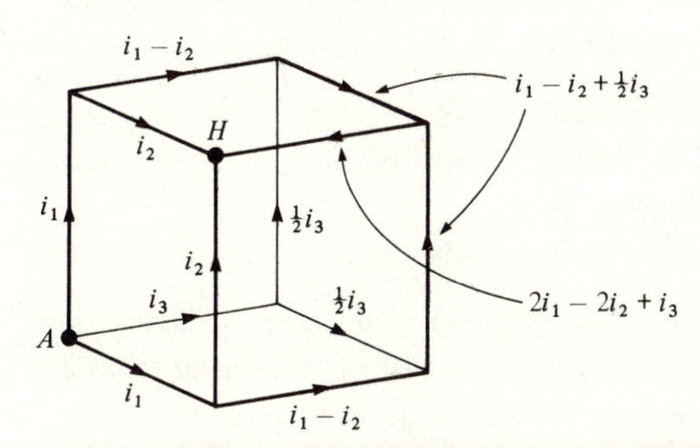

Use symmetry to keep the number of unknown currents to a minimum. The figure shows this for case (ii). The net voltage change round a closed loop is zero.

V13

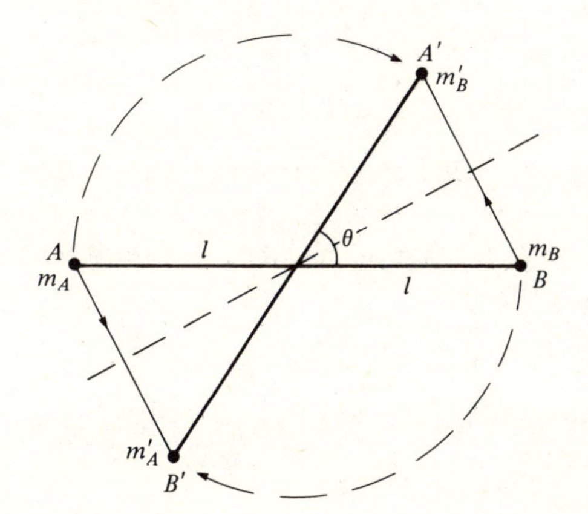

The fleas jump in directions making counterclockwise (say) angles $(\frac{1}{2}\pi - \frac{1}{2}\theta)$ with the initial direction of the hair. During the period in which they are in the air, the hair (because of the impulsive couple it receives) rotates clockwise about its centre through an angle $\pi - \theta$, so that both fleas land on the hair, but on the opposite ends from which they started. You will need to consider (in terms of the launch velocity v, the angle of take-off α and the length of the hair $2l$) the range of a flea's jump, its time in the air and the impulsive angular velocity acquired by the hair.

W Data handling

W1 Note that the reading of 9.28 m s^{-2} is clearly in error and should not be used in the calculation.

W2 The reading of 278 should probably be rejected. The remaining readings do not look to be Normally distributed and the best estimate is probably given by the inter-quartile range of the remaining 11 readings.

W3 (i) Reject the reading at 63.8.
(ii) Note that $\frac{2}{3}$ of the readings lie within ± 2 of the mean and that there are 14 readings being used.
(iii) Show that Student's t has a value of about 2.5 for 13 d.o.f.

W4 Obtain an initial value for $x = (p/p_0) - 1$ by ignoring the quadratic and cubic terms. Use this initial value to evaluate the higher terms and re-solve as a linear equation to obtain a more precise value for x. The two final values of x are -3.17×10^{-2} and -7.12×10^{-3}.

W5 (i) Plot h_2 versus d^3.
(iii) Plot $(h_2/h_1)^{1/2}$ versus $h_1^{1/2}$.

W6 Plot either $xy^{-1/2}$ versus $x^{1/2}$ or $(x/y)^{1/2}$ versus $x^{-1/2}$.

W7 Plot either x^2 versus (x/y) or xy versus (y/x).

W8 (i) The slope of the plot is -28.9 and the intercept when $1/T = 0$ is $\ln y = 1.11$.
(iii) Note that for large T, T^{-1} is small, and the exponential can be expanded with good accuracy.

W9 $X = 18.0 \pm 2.2$, $Y = 15.0 \pm 1.1$.

W10 (i) The expected distribution, if there were no correlation, is:

	Classical	None	Pop
Mathematics	17.5	10	22.5
None	24.5	14	31.5
English	28	16	36

This leads to a χ^2 of 12.3 for 4 d.o.f.

(ii) The expected distribution on the assumption of no correlation is:

	Music preference	No music preference
Academic preference	104	26
No academic preference	56	14

This leads to a χ^2 of 1.2 for 1 d.o.f.

W11 The 'votes' are not statistically independent and, once established, must be converted to a table of 'crossings out' before any χ^2 test is applied. This table is:

	Not X	Not Y	Not Z
Boys	50	35	65
Girls	25	40	40

The relevant values of χ^2 are (*a*) 9.0, (*b*) 4.3, and (*c*) 7.1, all for 2 d.o.f.

W12 (i) The distribution should be Binomial with $p = 0.314 \pm 0.021$. The expected number of times r golds are scored is calculated from

$$100\binom{5}{r}p^r(1-p)^{5-r}.$$

The cases $r=4$ and $r=5$ should be added since their expectations are small. χ^2 gives 0.4 for 4 d.o.f.
(ii) Expectation $= 31.4 \pm 4.6$. $\Phi[(40-31.4)/4.6] = 0.968$.
(iii) $\Phi[(0.360-0.314)/s] = 0.905$, where $s^2 = (0.021)^2 + (0.028)^2$.

W13 Total falls $= 377$. Total successful jumps $= 2825$. The average probability of falling $= 0.118$, and the expected falls at each fence are equal to this multiplied by the actual number of attempts, viz.

58.9, 51.6, 42.7, 37.0, 33.6, 29.7, 26.7, 23.2, 21.2, 19.0, 17.7, 15.9.

Comparison with the actual falls gives a χ^2 of 21.2.

ANSWERS

A Physical dimensions

A2 $410\,\mathrm{W\,m^{-1}\,K^{-1}}$.

A3 (i) Yes.
(ii) No, the B^2 term is incorrect. The correct expression is

$$S = \frac{1}{2(\epsilon_0\mu_0)^{1/2}}(\epsilon_0 E^2 + \mu_0^{-1}B^2).$$

(iii) Yes.

A4 (i) $v \propto \eta/\rho r$. (ii) $14\,\mathrm{m\,s^{-1}}$.

A5 $8^{1/2}$.

A6 $\mu = 40.7$. The actual value is $2\pi^5/15 = 40.8$.

A7 $\Sigma_q a_q (kT/F)^q l^{1-q}$, where q can take all values and a_q is a dimensionless constant dependent on q. The amplitude is independent of m.

A8 (i) $E \propto \rho R^5 t^{-2}$. (ii) $E \approx 8 \times 10^{13}\,\mathrm{J}$.

A9 $1.3 \times 10^{27}\,\mathrm{kg}$.

A10 $7.0 \times 10^{-25}\,\mathrm{s}$.

A11 (i) $T = (a^3/Gm_2)^{1/2} f(m_1/m_2)$, with f an arbitrary function. $T = (a^3/Gm_2)^{1/2} g(m_1/m_2)$, with g an arbitrary function, is equally valid. (ii) $T = 2^{1/2}T'$.

A12 (i) $f\left(\dfrac{\eta^2}{p\rho d^2}, \dfrac{\eta d}{\rho R}\right) = 0$.
(ii) It will flow at $\frac{1}{6}$ of the original volume flow rate when under $\frac{1}{18}$ of the original pressure difference.

A13 (i) $\lambda = A^{1/2} f(pA^{1/2}\sigma^{-1})$. The density ρ is not involved except through $p = \rho g h$. (ii) 6.05 mm (by interpolation). (iii) 0.16 m.

B Linear mechanics and statics

B1 (i) 55.9 mm. (ii) The same. (iii) For the optimum depth of water the centre of gravity of (beaker + water) lies in the water surface. Adding any further water necessarily raises the combined centre of gravity. The equation in the hint can be obtained either on this physical basis or by minimizing the expression for the height of the centre of gravity as a function of water depth.

B2 (i) $\frac{7}{2}M$. (ii) $4\frac{5}{6}$ m.

B3 (ii) 14.7 turns, i.e. $15\frac{1}{2}$ turns in practice.

B5 (i) $(MU \ln 3)/2F$. (ii) $(MU^2 \ln \frac{4}{3})/2F$.

B6 (ii) 10.8 m s^{-1}. $\mu > 0.050$.

B8 (*a*) (i) Backwards; the centre of gravity moves in the same direction as the force. (ii) Clockwise; corresponding to the wheels going backwards. (iii) Backwards and upwards; (i) and (ii) combined, but (i) is larger because of the gearing and wheel sizes.
(*b*) Increasingly negative - constant negative - decreasingly negative - same as originally.

B10 (i) $u \ln \left[\dfrac{M_2(M_1 + M_2)}{m_2(m_1 + M_2)}\right]$.

B13 (i) A at $(\frac{1}{2}L, \frac{1}{2}L)$ with velocity $(\frac{1}{2}V, \frac{1}{2}V)$; B at $(\frac{1}{2}L, \frac{3}{2}L)$ with velocity $(-\frac{1}{2}V, \frac{1}{2}V)$.
(ii) A at $(L, 50L)$ with velocity $(0, V)$; B at $(0, 50L)$, at rest.

C Circular and rotational motion

C1 $\dfrac{\omega}{\omega_0} = \left[\dfrac{f_0(f-1)}{f(f_0-1)}\right]^{1/2}$.

C2 355 atmospheres.

C4 (i) $\omega^2 = g/l$. (ii) K.E $= \frac{1}{2}J\omega =$ P.E., the P.E. being measured from the at rest position. (iii) $W = J(\omega - \omega_0)$.

C5 (ii) $\omega = 20.9$ rad s^{-1}.

C7 $\frac{2}{3}\pi$.

C8 (i) $\omega_0 = (8g/3a)^{1/2}$.

C9 (i) $Mr(V-\omega r+2g\omega^{-1})$. (iii) $(V^2+8gr)/(2V^2+8gr)$.

C10 (i) $I=\frac{1}{3}ma^2$. (ii) 8.2×10^{-6} s.

C11 (*a*) $R=P[(3y/2l)-1]$. (*b*) $y=\frac{7}{10}l$ for $R=0$.

C12 (i) $J=3^{1/2}mR\omega_0/2$. (ii) Final K.E. $=mR^2\omega_0^2/8=\frac{1}{4}$ initial K.E.

C13 $u=\frac{2}{3}(gh/3)^{1/2}$.

D Gravitation and circular orbits

D1 3.9×10^8 m.

D2 1.4×10^{10} m.

D3 0.32 s.

D4 1.8×10^{-9} kg.

D5 (i) 8.0×10^{-6} per revolution. (ii) Increases.

D6 (i) λ^{-1}, larger. (ii) λ^2, smaller. (iii) λ^2, larger, i.e. less negative.

D7 If the force is kr^{-3}, $E=0$ and $J=(km)^{1/2}$ for any r. Neutral equilibrium in any orbit.

D9 3.6×10^{-5} m s^{-1}.

D10

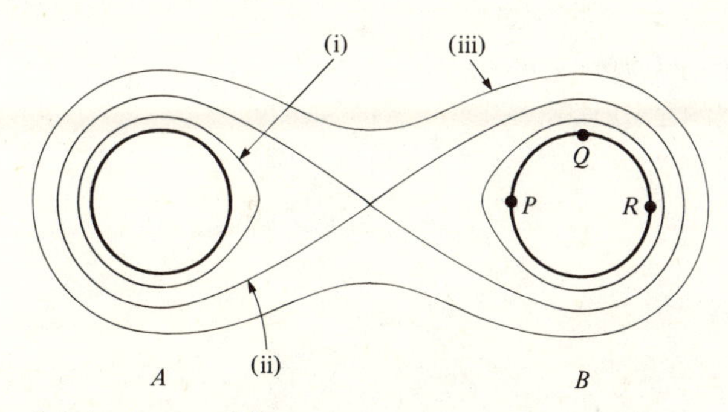

(iv) $(20GM/21r)^{1/2}$ starting from point R on star B.

D11 (i) 2.1×10^{-6}. (ii) Yes, the effect is proportional to M/R which is a factor of about 3000 smaller for the Earth than the Sun.

D12 (ii) 2.0×10^{-3} radians.

D13 (i) $v_A=m_X[Gd^{-1}(m_A+m_X)^{-1}]^{1/2}$, $\omega=[G(m_A+m_X)d^{-3}]^{1/2}$.
(iii) Probability = 0.4.

E Simple harmonic motion

E1 (*a*) (i) 0. (ii) $a = 0.02$ m, $\omega = 3.14$ rad s^{-1}, $\epsilon = \frac{1}{2}\pi$.
(*b*) (i) 2.9×10^{-13} m. (ii) 547 m s^{-1}.

E2 (i) 5.0. (ii) 7.3.

E3 $\pi x_0^2 (mk)^{1/2}/2fl$.

E4 (i) 9.9 rad s^{-1}. (ii) 39.2 N. (iii) 0.21 m.

E5 (*a*) $2\pi(l/g)^{1/2}$. (*b*) (ii) $\frac{1}{2}D$, $2\pi(2D/g)^{1/2}$.

E6

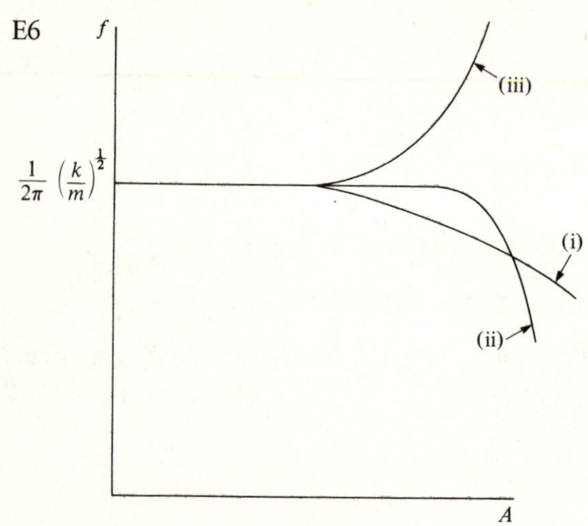

E7 (i) 1.20 s. (ii) 60 r.p.m.

E8 $2\pi(l/2g)^{1/2} = 1.42$ s

E9 7.60 s.

E10

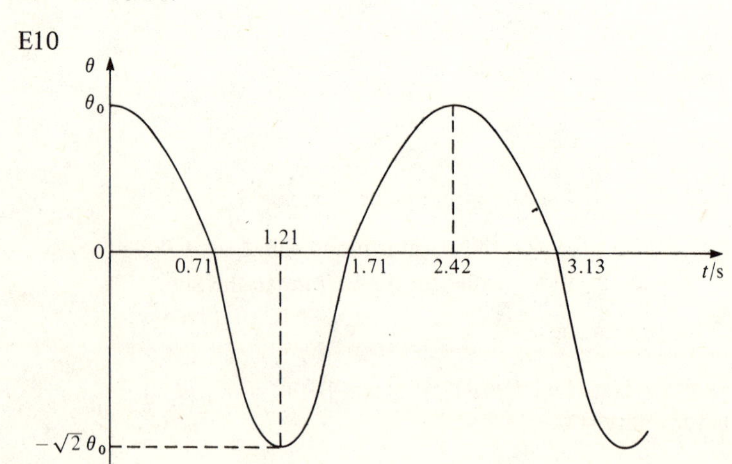

E11 (*a*) 2.25.

(*b*)

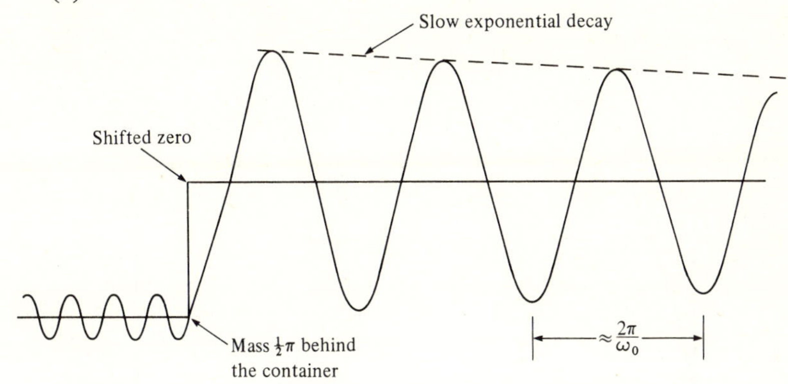

(*d*) $B/A \approx b/m\omega$, $\phi \approx -\frac{1}{2}\pi$.

E12 (i) $M/(M+m)$. (ii) No change.

E13 (*a*) (i) $M\ddot{x} = -kx$. (ii) $M\ddot{x} = -4kx$.

(*b*) and (*c*)

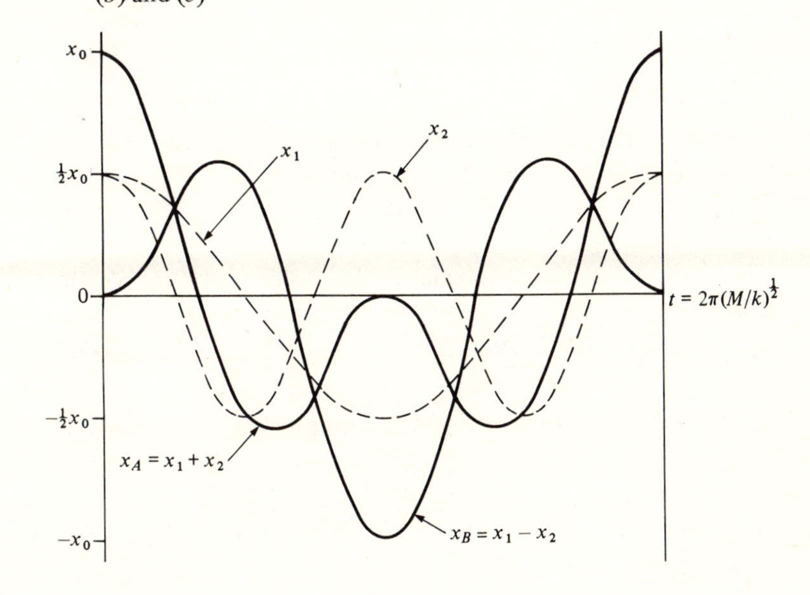

F Waves

F1 $0.25\,\mathrm{m\,s^{-1}}$.

F2 (*a*) (i) $11.0\,\mathrm{m\,s^{-1}}$. (ii) $24.5\,\mathrm{m\,s^{-1}}$.

(*b*) T.

F3 (*a*) (i) 0.10 m. (ii) Increase by 2%.
(*b*) 178 Hz.

F4 (*a*) $345\,\mathrm{m\,s^{-1}}$.
(*b*) (i) 583 Hz. (ii) 1710 Hz.

F5 (i)

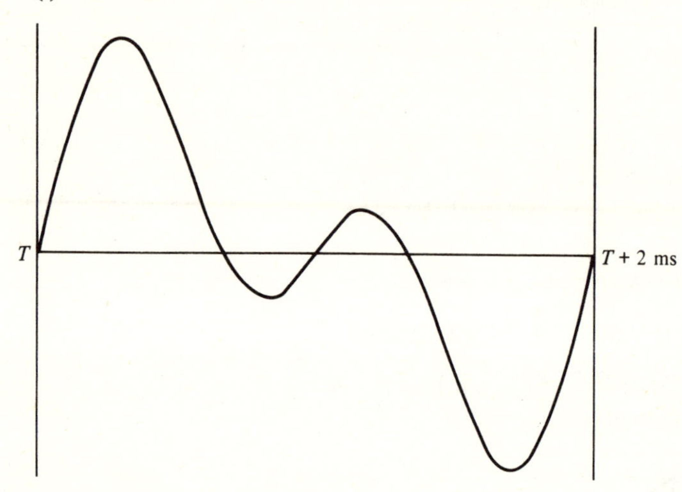

(ii) 44 mm.

F6 A descending tone falling towards a frequency $f_0 = v/2a$;
$f = f_0 t(t^2 - t_0^2)^{-1/2}$, where $t_0 = 2D/v$ and $t > t_0$.

F7 $0.32\,\mathrm{m\,s^{-1}}$.

F8 (*a*) (i) 500 Hz. (ii) 538 Hz and 467 Hz. (iii) 546 Hz and 460 Hz.
(*b*) (i) $+8.3 \times 10^{-2}$ nm. (ii) $\pm 3.3 \times 10^{-3}$ nm.
(*c*) $f' = f\left(1 + \frac{\pi}{70} \sin \frac{2\pi t}{3}\right)$.
(*d*) None, except for a totally negligible mistuning of the radio.

F9 121 Hz.

F10 (i) 196.0 Hz. (ii) $25.7\,\mathrm{m\,s^{-1}}$. (iii) 76.3 s.

F11 $3.3 \times 10^{-2} \sin^2 2\theta$.

F12 $y = \frac{1}{4}(2n + 1)\lambda - (2)^{-1/2}L$, for integral n large enough to make y positive.

F13 $a = 3.23, b = 8.08, c = 12.62, d = 16.66$; $2\,\mathrm{m\,s^{-1}}$ in the negative x-direction; amplitude = 25 cm.

G Geometrical optics

G1 (i) 300 mm above lens. (ii) $1200(200-h)$ mm in diameter.

G3 12.5 mm.

G4 (*a*) (i) 1.5 mm. (ii) 1.526 mm.
(*b*) 1.5 mm and 4.0 mm.

G5 Not quite (not dependent on the 200 mm value). Slit images 1.12 mm wide with centres separated by 1.19 mm.

G6 (ii) $A = -5.00°$ (prism reversed), $A' = 10.00°$.

G7 (i) Eyepiece tube has to be pulled out by $12.5 - 10.0 = 2.5$ mm.
(ii) Image is 17.5 mm in diameter.

G8 (i) 2.026 mm. (ii) 1043. (iii) 1.03 mm. (iv) 2.5×10^{-4} mm.

G9 (i) The 40 mm and 45 mm lenses in contact as the objective, the 30 mm lens as eyepiece. Magnification 12. (ii) The 30 mm and 40 mm lenses, either way round. Magnification 16.7, taking the least distance of distinct vision as 250 mm.

G10 (ii) and (iii)

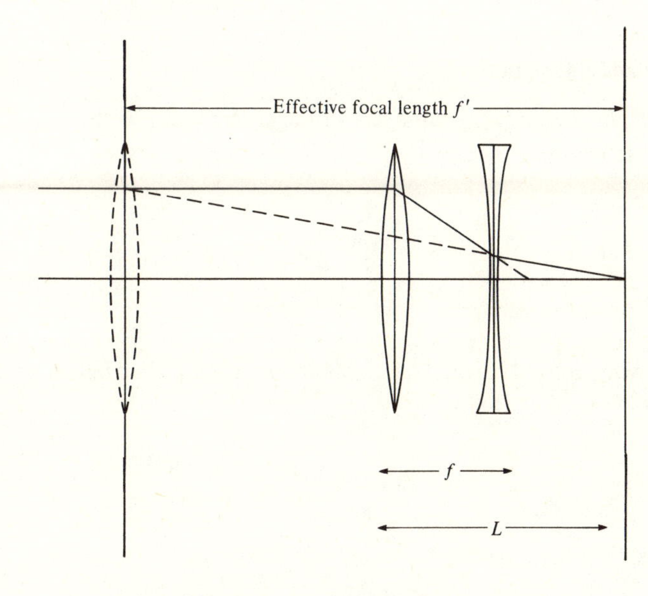

$f' = 210.9$ mm, $L = 104.4$ mm.

G11 (i) $d = 1510$ m. (ii) Temperature falls.

G12 (*a*)

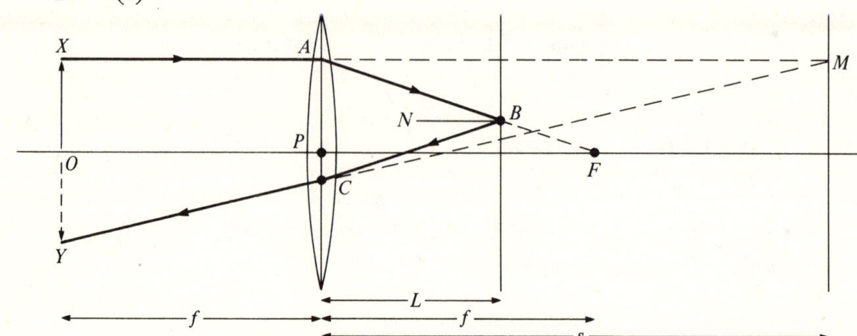

(1) If object and image positions coincide, then $OP = f$, since then the rays from O strike the mirror normally and are directly reversed.
(2) Make $OY = -OX$ and join A to F to determine B.
(3) Make $ABN = NBC$ and so determine C.
(4) Produce YC back to meet XA produced at M, the 'plane' of the concave mirror.

G13 (*b*) (i) 0 and $\frac{1}{2}$ revolution: inverted image. $\frac{1}{4}$ revolution: erect image.
(ii) Arrow rotates about the axial ray with twice the angular frequency of the prism.

H Interference and diffraction

H1 (*a*) 1.25×10^{-5} m. Towards the same side of the centre line as is the film. Use white light (see part (*b*)).
(*b*)

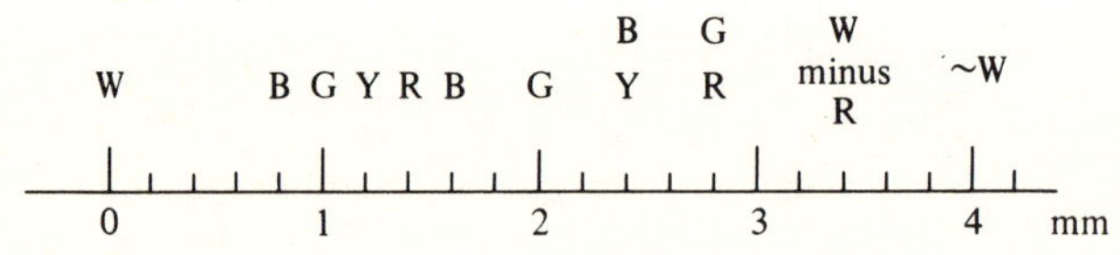

H2 (*a*) Maxima when $RT = 12.6$ m, 37.9 m. Minimum when $RT = 18.9$ m.
(*b*) (i) 4. (ii) 4.

H3 3.7 m.

H4 (i) 10^{-5} m. (ii) Dark.

H5 $1.99 \times 10^{-5}\,\mathrm{K}^{-1}$ or $8.1 \times 10^{-6}\,\mathrm{K}^{-1}$.

H6 (i) 55.9°. (ii) 11.9°, 38.4°.

H7 (i) $\sin\theta = (n + \frac{1}{2})\lambda/d$, where θ is the angle between the direction and the normal to the line of transmitters. (ii) $\sin\theta = m\lambda/2d$, i.e. as in (i) for m odd, but with additional directions for m even.

H8 (*a*) About 2.6 km for headlamps 1.5 m apart. (*b*) About 800.

H9 (*a*) (i) Fringes 1 m apart centred on the symmetry line. (ii) Pattern shifted 0.25 m to the left. (iii) Pattern moves left at a rate of $1\ \mathrm{m\,s^{-1}}$. (*b*) The intensity does not vary with position, but the resultant polarization does as shown qualitatively below.

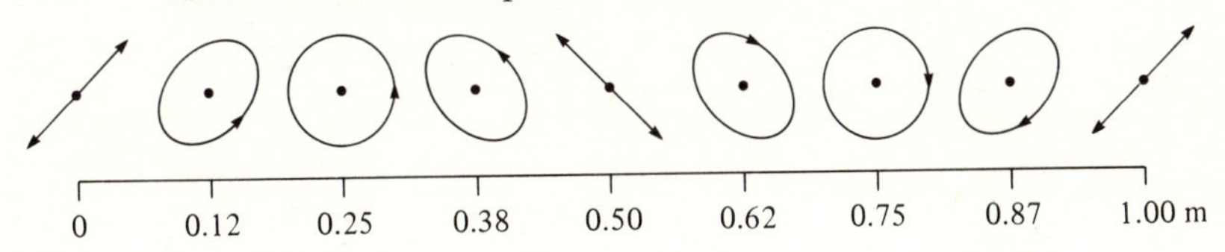

H10 Transmitter is 26.6° South of West. Bus speed $= 11.2\ \mathrm{m\,s^{-1}}$.

H11 $\alpha = 6.4 \times 10^{-5}$ rad.

H12 $56(2n+1)\,\mu\mathrm{m},\ n = 0, 1, 2, \ldots$.

H13 (*b*) 99 m.

I Structure and properties of solids

I1 6.7×10^{-8} kg.

I2 $\theta_0 = 14.2°$.

I3 $\lambda = (1.42 \times 10^{-10}/n)$ m, where $n = 1, 2, 3, \ldots$.

I4 (i) $x_0 = (2B/A)^{1/6}$. (ii) $18(A^{13}/2B^7)^{1/6}$. (iii) $(13/7)^{1/6} - 1$.

I5

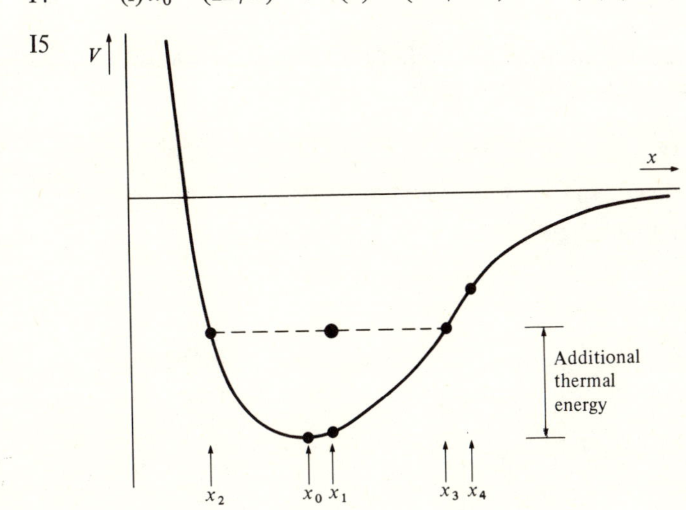

(i) When the temperature increases and additional thermal energy is added, the mean separation x_1 ($\approx\frac{1}{2}(x_2+x_3)$) becomes greater than x_0 because the curve is not symmetric about x_0.
(ii) The Young modulus at any particular temperature is $\mathrm{d}^2V/\mathrm{d}x^2$ at the mean separation for that temperature. As x increases (above x_0), $\mathrm{d}^2V/\mathrm{d}x^2$ decreases (towards zero at $x=x_4$).

I6

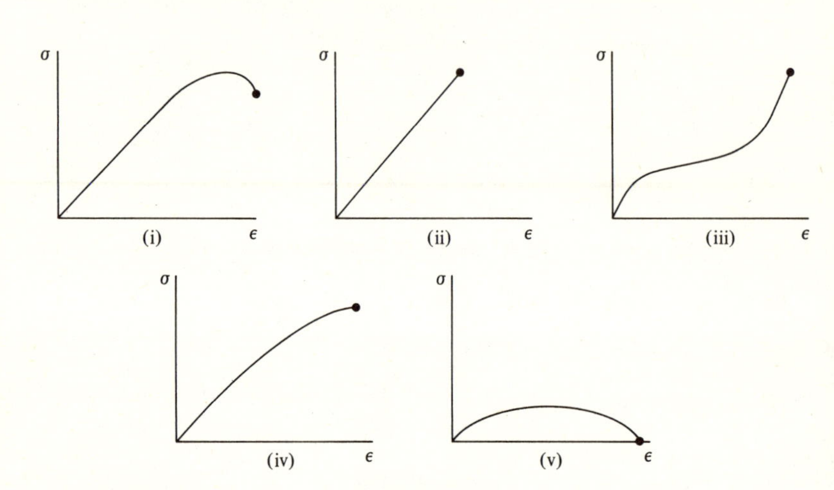

I7 $x_0=0.076$.

I8 0.37 m.

I9 $\Delta R=R^3\omega^2\rho/E$.

I10 $$x=\frac{L\Delta T(E_1\alpha_1-E_2\alpha_2)}{E_1+E_2}.$$

I11 (i) $v=(2gh)^{1/2}\cos[(EA/ml)^{1/2}t]$. (ii) $\tau=\pi/\omega=5.2\times10^{-4}$ s. (iii) 5.4×10^{4} N. (iv) 3.0×10^{-3}.

I12 (*a*) 4.5 K. (*b*) 0.13, several times the practical value of 2–3%.

I13 $v_0=20\,\mathrm{m\,s^{-1}}$. The recoverable stored energy is $\frac{1}{2}Mv_0^2$ in all cases, and hence v_r always equals v_0.

J Properties of liquids

J1 (i) Falls by $M_\mathrm{s}A^{-1}(\rho_\mathrm{W}^{-1}-\rho_\mathrm{s}^{-1})$. (ii) No change. (iii) Falls by $M_\mathrm{s}A^{-1}(\rho_\mathrm{W}^{-1}-\rho_\mathrm{s}^{-1})-M_\mathrm{T}A^{-1}(\rho_\mathrm{T}^{-1}-\rho_\mathrm{W}^{-1})$, i.e. by less than in (i).

J3 (i) $\frac{1}{4}+\frac{3}{4}(\rho'/\rho)$.

J5 20 mN m^{-1}.

J6 (i) 4.8 mm. (ii) 0.20 g. (iii) No, the pressures on the horizontal faces of the inner tube are equal (both atmospheric).

J7

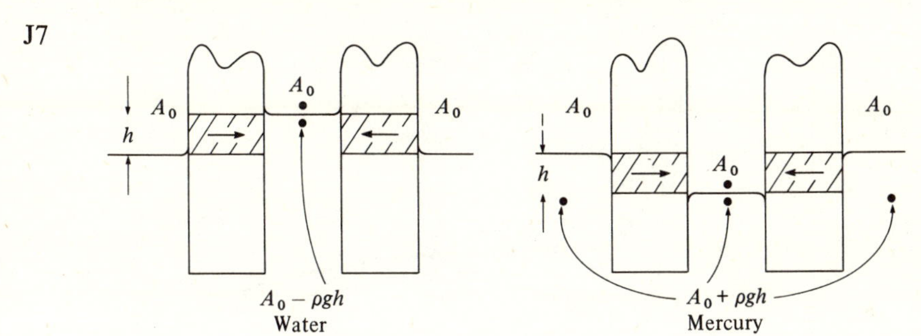

There are net inward forces acting on the shaded areas in both cases.

J9 (i) $l = 13(3)^{1/2}\epsilon/6\rho r^3$. (ii) Parallel to the surface.

J10 $\omega_0 \exp(-\pi\eta a^2 t/Md)$.

J11 (*b*) $8l\eta a^2 b^2 \ln 2/[\rho g r^4(a^2 + b^2)]$.

J12 296 s.

J13 $w\rho g d^3 \sin\theta/3\eta$.

K Properties of gases

K1 (i) 168 J. (ii) 149 J.

K2 $N_0/8$.

K3 $2.4 \times 10^{-3}\,\text{m}^3$.

K4 $\frac{1}{8}p \ln 5 = 0.20p$.

K5 $T_f = \frac{7}{5}T$.

K6 $2.5 \times 10^{-6}\,\text{m}^2$.

K7 5400 m.

K8 11.2 kW.

K9 (i) $l = \dfrac{V_0}{A}\left[\left(\dfrac{p_1}{p_0}\right)^{1/\gamma} - 1\right]$.

(ii) $\dfrac{p_1 V_0}{\gamma - 1}\left[1 - \left(\dfrac{p_0}{p_1}\right)^{(\gamma-1)/\gamma}\right] - p_0 V_0\left[\left(\dfrac{p_1}{p_0}\right)^{1/\gamma} - 1\right]$.

K10 $(a)\ p_0\left[\left(\frac{m_{\mathrm{He}}}{m_{\mathrm{A}}}\right)\left(\frac{T(x)}{T(0)}\right)\right]^{\gamma/(\gamma-1)}$.

K12 (iv)

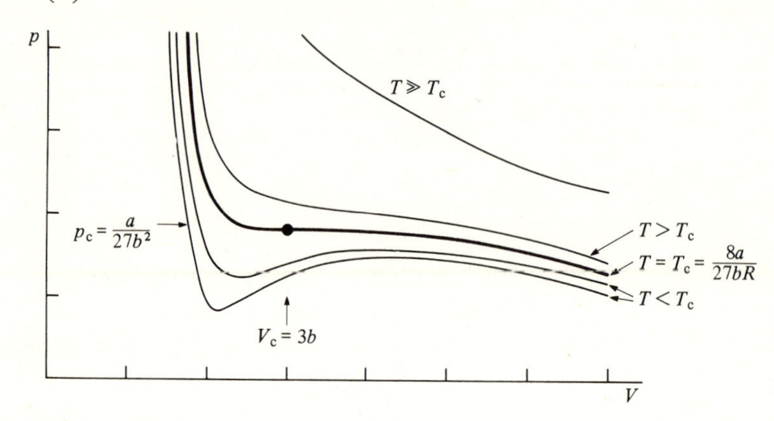

K13 (ii) 7×10^{-6}.

L Kinetic theory and statistical physics

L1 Divide by $T_1/T_2(=15)$.

L2 (i) 50 N. (ii) $4L^2 \ln 2/d\bar{c} \approx 79$ s. (iii) (i) requires $\overline{c^2}$ which is proportional to T without approximation. In (ii) $\bar{c}$ is approximated by $c_{\mathrm{r.m.s.}}$. The actual velocities will have a distribution; the faster ones collide with the walls more often and are therefore more likely to escape. Consequently the average velocity of those remaining falls and results in a decrease of $d\bar{c}/4L^2$ with time.

L3 0.23 s.

L4 (i) $\phi = p(9k/2\pi mT)^{1/2}\Delta T$. (ii) Independent of d. (iii) $\phi' = \phi/(N+1)$.

L5 (iii) $\Delta T = \frac{mv_0^2}{3k}\left(\frac{x_0^2}{x^2} - 1\right)$.

L6 $F \approx IA/c_1 = 7.1 \times 10^{-6}$ N.

L7 (a) 32/70. (b) (i) $R \ln 4 = 11.5\ \mathrm{J\,K^{-1}}$. (ii) Since the change is irreversible, the heat absorbed is less than $300R \ln 4 = 3452$ J.

L8 (i) Resistor, $\Delta S = 0$. Water, $\Delta S = 3.3\ \mathrm{J\,K^{-1}}$. (ii) $2.42\ \mathrm{J\,K^{-1}}$.

L9 (i) $\exp(7.8 \times 10^{24})$. (ii) $\exp(4.3 \times 10^{22})$.

L10 $15.8\,\mathrm{J\,K^{-1}}$.

L11 (i) $R \ln 2$ with $\Delta Q = RT \ln 2$. (ii) $R \ln 2$ with the heat absorbed less than $RT \ln 2$ since the process is irreversible.

L12 75°C, assuming that the specific latent heat is independent of temperature, that the same fraction of molecules in any particular energy range escape at all temperatures, and that evaporated molecules do not return.

L13 $C(T_1 + T_2) - 2C(T_1 T_2)^{1/2}$.

M Heat transfer

M1 (i) $T = T_0[1 - (\mathrm{d}V^2/\lambda a A)]^{-1}$. (ii) The thermistor overheats and burns out.

M2 91.4 hours.

M3 117 days, assuming both start at the same temperature.

M4 (i) $T = \frac{1}{3}(2T_1 + T_0)$. (ii) $(6lC \ln 2)/\pi d^2\lambda$.

M5 $Ha^2/6\lambda = 0.20\,\mathrm{K}$.

M7 $[C(T_1 - T_0) \ln 2]/P$.

M8 234 K.

M9 1600 K.

M10 (i) 40.4 °C, 63.9 °C, 83.3 °C. (ii) $\phi' = \frac{1}{4}\phi$.

M11 The observed output is about 21 times the expected output in the visible region. The fluorescent powder absorbs a significant part of the ultraviolet light emitted by the discharge and re-emits it at longer wavelengths, some of it in the visible range.

M12 2.50 kW.

N Electrostatics

N1 $4\epsilon_0 A V^2/d$ increase.

N2 (i) $(-\epsilon_0 A V^2/2x^2)\,\mathrm{d}x$, (ii) $(+\epsilon_0 A V^2/2x^2)\,\mathrm{d}x$. (iii) In (i) the charge on the plates decreases, and the energy of the battery is increased by $V\,\mathrm{d}Q = V^2\,\mathrm{d}C = (\epsilon_0 A V^2/x^2)\,\mathrm{d}x$. (iv) $\epsilon_0 A V^2/2x^2$.

N3 (i) 50 V, 20 V. (ii) No changes. (iii) Both $28\frac{4}{7}$ V. (iv) Both 0. (v) 70 V, 40 V.

N4 (i) -2.64×10^{-7} C, -1.32×10^{-7} C. (ii) 2250 V.

N5 1.1×10^{-7} J.

N7 $4\pi\epsilon_0 a V^2$.

N9 (i) 4.1×10^{-9} C. (ii) No, the electric field strength at the surface of the drop would be 3.7×10^7 V m^{-1}.

N10 No component of frequency ω if $2\alpha Q_0 = d_0$.

N12 (i) $\frac{1}{4}bE_0$. (ii) $27\pi\epsilon_0 E_0^2 b^3/128$.

N13 (*a*) $W = 2\pi a^3 \sigma^2/\epsilon_0$. (*b*) 9.9×10^6 N m^{-2}. (*c*) If ionization is the limiting factor, use one big sphere. If the tensile strength of the material is the limiting factor, it does not matter whether one large sphere or several smaller ones are used.

O Direct currents

O1 1.01 kW.

O2 (*a*) 1.48×10^{-4} m s^{-1}. (*b*) 2.1×10^{-8} Ω m.

O3 (i) 60 Ω at 2 km from A. (ii) 37%.

O4 (i) 0.47 W. (ii) 52.0 kW calculated from

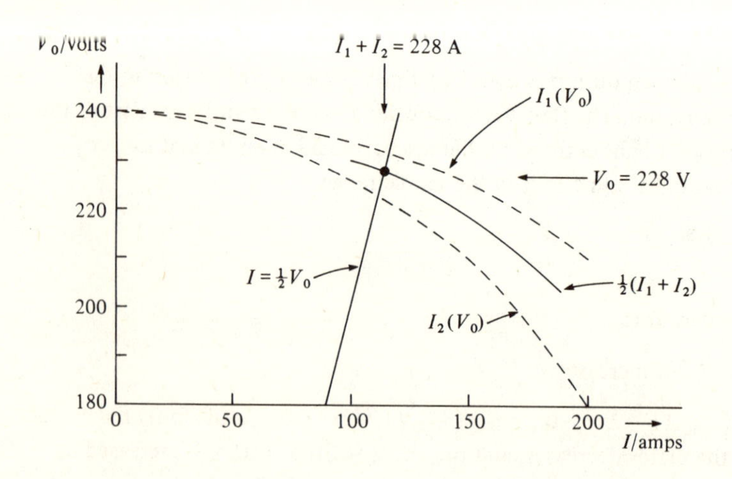

O5 457 A.

O6 A, who obtained only 8 W of lighting. B obtained 32 W. Clearly B was the brighter.

O7 (i) 2.5 mΩ in parallel. (ii) 4.8 MΩ in series. (iii) 480 MW.

O8 (*a*) (i) $r=R$. (ii) r as large as possible.
(*b*) $R'=0.022\ \Omega$.

O10 $rV[2R(R+r)+G(2R+r)]^{-1}=3.3\times10^{-7}$ A.

O11 (*a*) (i) 6.8 Ω. (ii) 5 Ω. (iii) 3.37 Ω.
(*b*) 2 A.
(*c*) 90 Ω.

O12 $a=1\ \Omega, b=3\ \Omega, c=4\ \Omega, d=5\ \Omega, e=2\ \Omega, f=6\ \Omega$.

O13 (*c*) The curve is asymptotic to the dashed line for large positive values of V.

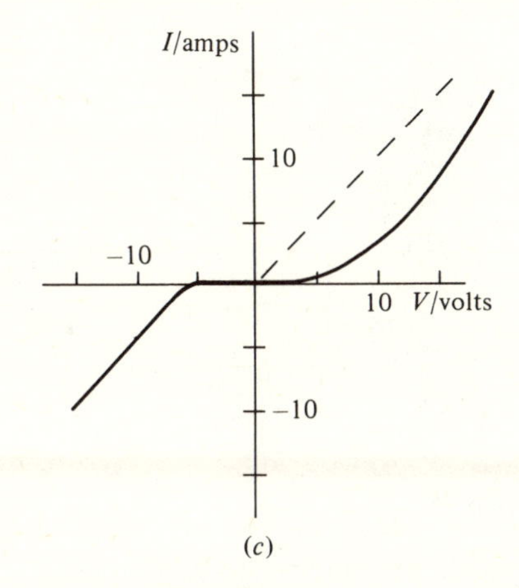

(*c*)

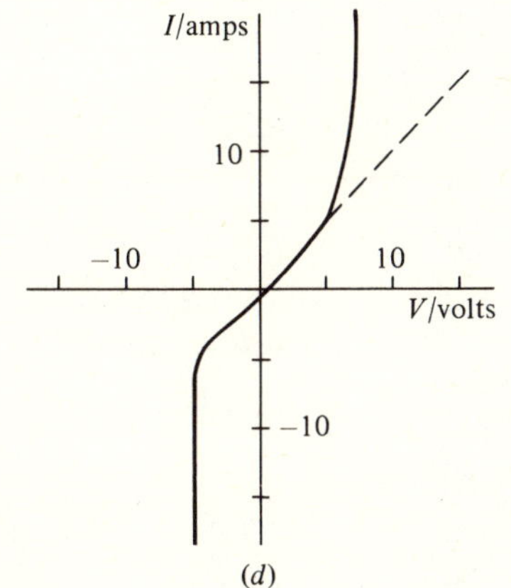

(*d*)

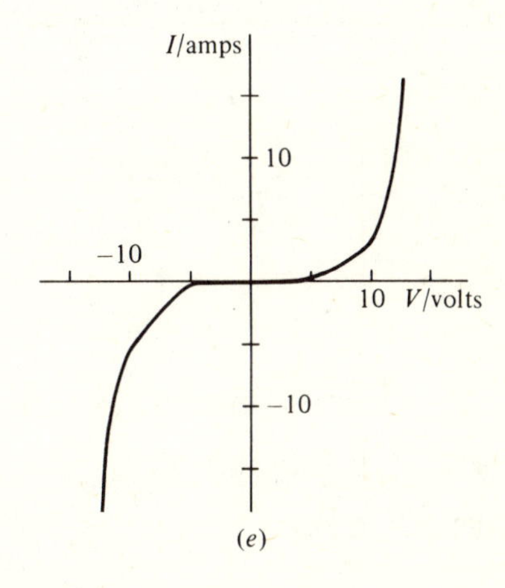

(*e*)

P Non-steady currents

P1 4.3×10^{-6} m.

P2 8.33 V.

P3

(*a*)

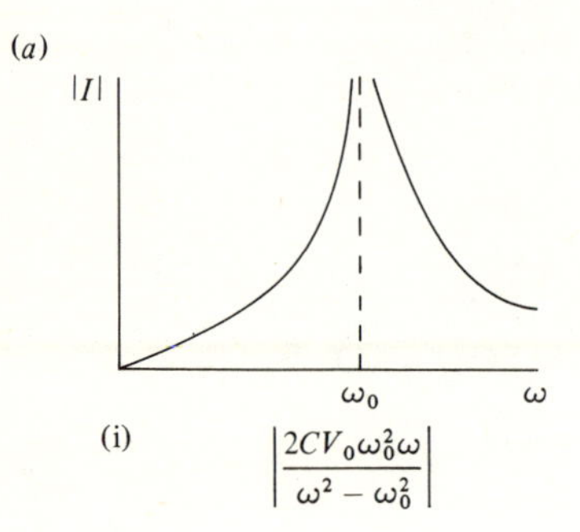

(i) $\left|\dfrac{2CV_0\omega_0^2\omega}{\omega^2 - \omega_0^2}\right|$

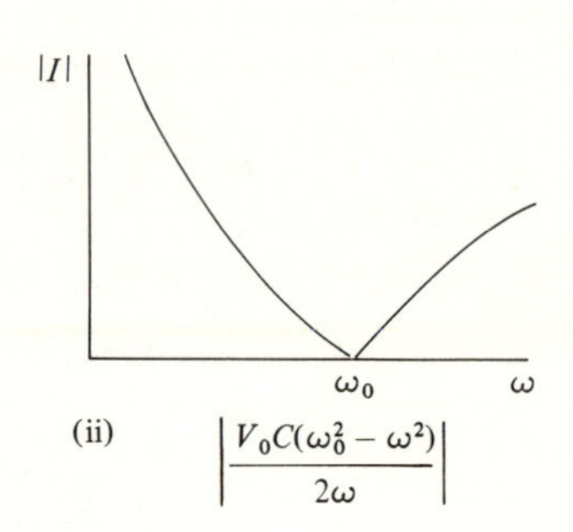

(ii) $\left|\dfrac{V_0C(\omega_0^2 - \omega^2)}{2\omega}\right|$

(*b*)

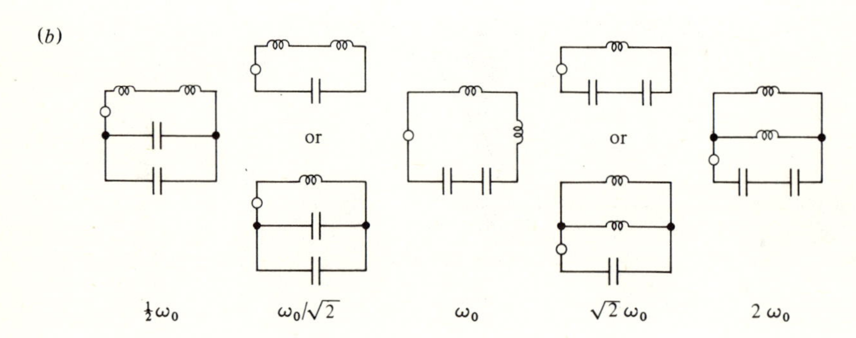

P4 (i) 192 Ω, 150 W. (ii) 1.06 H, 75 W.

P6 The waveforms drawn are only qualitative and not to scale.

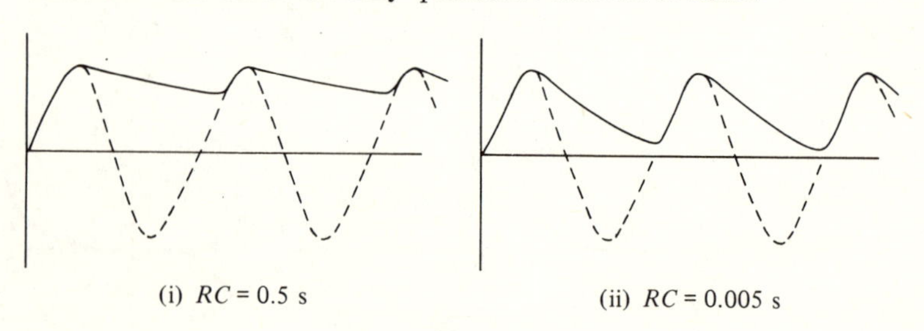

(i) $RC = 0.5$ s (ii) $RC = 0.005$ s

(iii) 0.39 V.

P7 0.55 H, 18.4 μF.

P8 0.2 s.

P9 (ii)

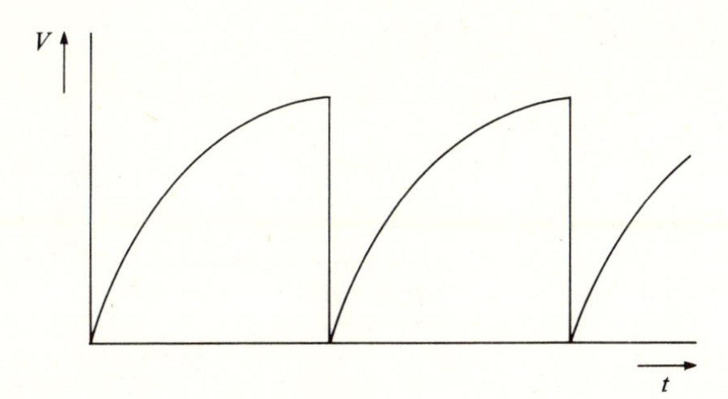

$RC > 0.5$ s; $V_0 > 50\, V_1$.

P10 (*a*) $[2 + (10/7) \exp(-t/0.07)]$ mA.

(*b*)

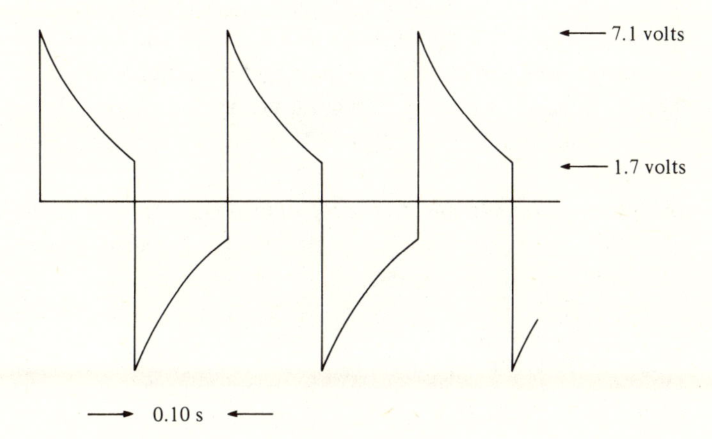

P11 (*a*) $R = 500\,\mathrm{k}\Omega, C = 2\,\mu\mathrm{F}$. (*b*) $\tau = RC = 1$ s.

P12

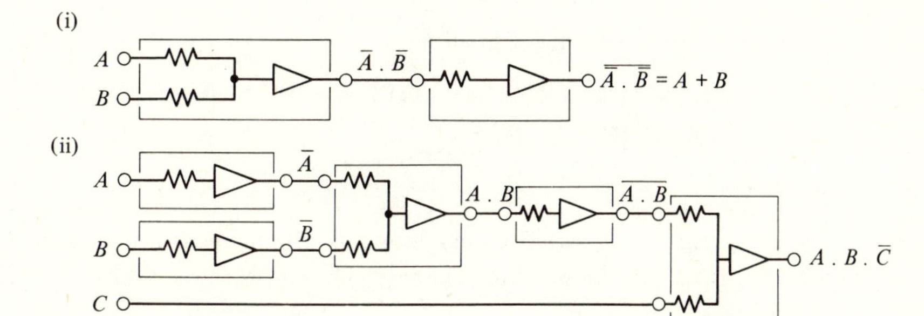

P13 (i)

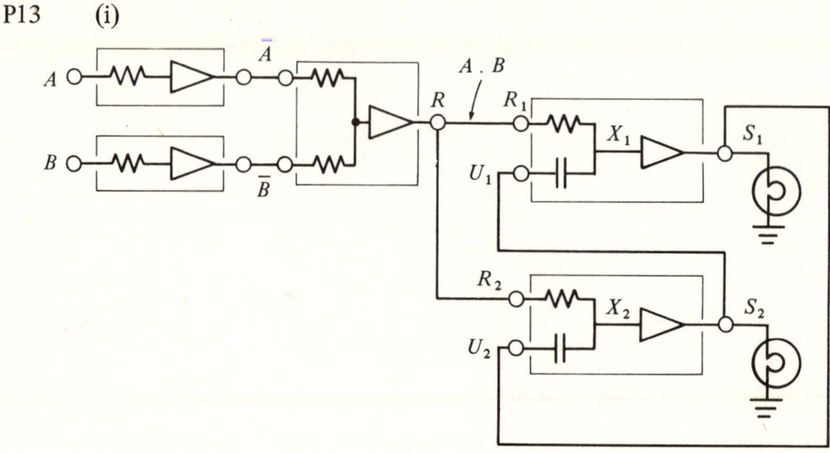

R will be at 6 V only if A and B are both at 6 V. With this condition satisfied, if a positive pulse of 6 V amplitude appears at U_1, the end-of-the-pulse drop to 0 V momentarily sets X_1 to 0 V and hence S_1 to 6 V. X_1 recovers to 6 V with time constant $\tau_1 = R_0C_0$ and S_1 drops to 0 V. Thus S_1 transmits a positive pulse of 6 V amplitude to U_2. The process then repeats, turning S_1 and S_2 to 6 V alternately and thus lighting the lamps.

(ii) For example, add $3R_0$ to the R_1 input making τ_1 become $4R_0C_0$, or add $\frac{1}{3}C_0$ to the U_2 input making τ_2 become $\frac{1}{4}R_0C_0$. Many other combinations of changes are possible.

Q Magnetic fields and currents

Q2 (ii) $1.59\ V < V_0 < 9.55$ V.

Q4 (i) 1.7×10^9 A. (ii) Anticlockwise as viewed from above the North Pole.

Q5 (i) $10^{-4}\,\mathrm{m\,s^{-1}}$. (ii) The edge $x = 0, y = +a$.

Q6 (i) $r = mv/Be$, $T = 2\pi m/Be$. (ii) 2.4×10^{-28} kg.

Q7 $(2\pi mv \cos\theta)/Be \approx 3.6$ m for small θ.

Q8 (i) 3.2×10^{-6} N towards the wire. (ii) 1.32×10^{-6} N away from the centre of the rectangle.

Q9 3.3 MW.

Q10 (i) $k_0 = \mu_0 i I a^2/2R$. (ii) The small loop sets with its plane orthogonal to that of the large loop. (iii) The small loop turns through 108.6° from its initial position.

Q11 The situation is unstable and the coil executes (damped) oscillations about an orientation directly reversed from its original one. When the oscillations have become sufficiently small, their angular frequency is $(\mu_0 i l/mr)^{1/2}$.

Q12 (i) Opposite senses.

Q13 (i) $C_X = M^2/L^3$ clockwise (say), $C_Y = 2M^2/L^3$ clockwise.
(ii) Nothing would happen; the force $F_X (= 3M^2/L^4)$ acting on X because of the field due to Y, and its reaction on the rod produce a couple $C = 3M^2/L^3$ anticlockwise, exactly cancelling $C_X + C_Y$.

R Electromagnetism

R1 0.13 MPa.

R2 An approximately sinusoidal voltage of amplitude 9.1 mV and period 12 hours, with the North bank positive when the tide is coming in.

R3 (i) 0.96 V from port wing tip (−ve) to starboard wing tip (+ve).
(ii) None. (iii) 48 mV from top (−ve) to bottom (+ve). (iv) 0.73 V from port wing tip (−ve) to starboard wing tip (+ve), and 23 mV from bottom (−ve) to top (+ve).

R4 $\frac{1}{2}B\omega L(L-2x)$.

R5 (i) 226 V. (ii) 3.5×10^7 A.

R6 1.75 mm s^{-1}.

R8 (i) 1.43 T. (ii) 4.0×10^{-4} N m. (iii) $I = kn(an^2 + b\omega^2 n^4)^{-1/2}$, where k, a and b are constants. Thus $I \propto n^{-1}$ for large n.

R9 (i) $LI_0\omega \cos \epsilon + RI_0 \sin \epsilon = \omega\Phi_0$ and the equation given.
(ii) $P = \frac{1}{2}Rv^2\Phi_0^2(L^2v^2 + R^2r^2)^{-1}$.

R11 (i) 0. (ii) $2\pi^2 a^2 R\dot{B} \sin\theta\,[\theta(2\pi - \theta)r + 4\pi^2 R]^{-1}$.

R13 $$\frac{B^2 vd^2}{2\rho(d+x)}\left(1 - \frac{\mu_0 vd}{4\pi\rho x(d+x)}\right).$$

S Nuclei

S1 (*a*) 5.7×10^{-14} m. (*b*) About 17 MeV.

S2 (*a*) 2.7×10^{-13} kg. (*b*) 2.6×10^{-2} mol.

S3 Half-lives: A, 0.36 d; B, 4.8 d. Initial numbers, 2.6×10^7 of each.

S5 40.7 days to cover 100 miles seems every reason to complain.

S6 (i) 9.5×10^{17}.
(ii)

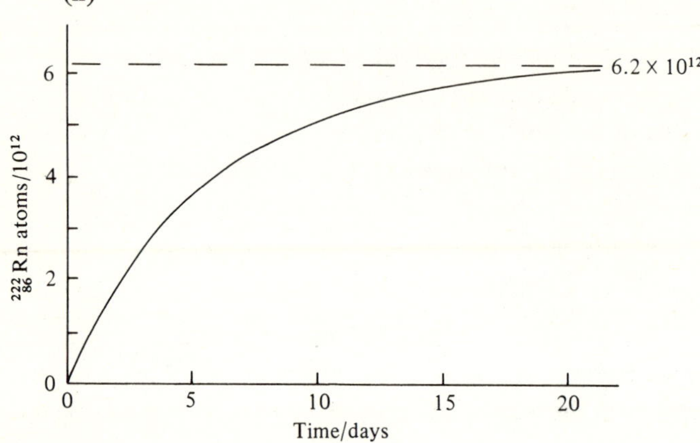

(iii) $6.5 \times 10^9 \ln [1 + (A_{Pb}/A_U)]$.

S7 (i) The equilibrium value of 3.4×10^5 atoms. (ii) The same.
(iii)

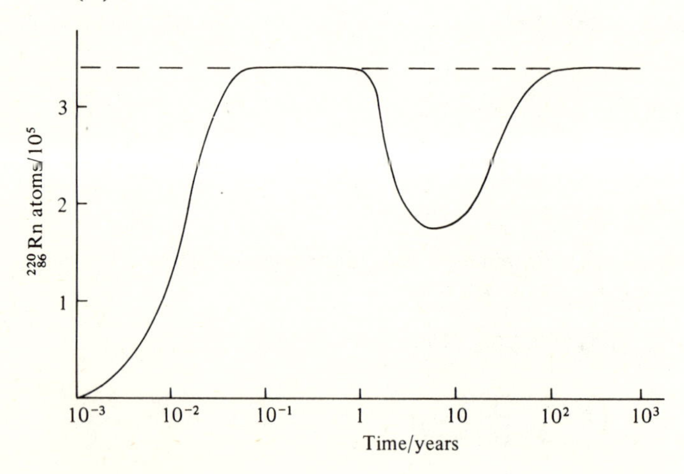

S8 (i) 10.087 MeV. (ii) Not possible; charge not conserved. (iii) 3.367 MeV.

S9 4.248 MeV.

S10 (i) $^{13}_{7}N$, $^{13}_{6}C$, $^{14}_{7}N$, $^{15}_{8}O$, $^{15}_{7}N$, $^{12}_{6}C$, $^{16}_{8}O$, $^{17}_{9}F$, $^{17}_{8}O$, $^{14}_{7}N$.

(ii) $4p \rightarrow \alpha + 2e^+ + 2\nu +$ energy, for both. All carbon, nitrogen and oxygen isotopes are catalysts.
(iii) 5.9×10^{11} J.

S12 (i) $A_s = 80, B = 1.15 \times 10^{-10}$ J. (ii) $A_f = 115$. (iii) No. Spontaneous fission requires $C(A_f) < C(A_f/2)$; but $C(A_s) > C(A)$ for any other A, implying that $A_f \not> 2A_s$. Here $C(A) \equiv A^{-1}B(A)$. (iv) Yes, for all nuclei with $A < \frac{1}{2}A_f \approx 57$.

T Electrons, photons and atoms

T1 4.1 mm.

T2 Measured in units of the electronic charge, quarks have charges $-\frac{2}{3}$ or $+\frac{1}{3}$. Possible charges of $+1$ cannot be established because of the presence of electrons.

T3 (i) 6.64×10^{-34} J s. (ii) 4.77 eV.

T4 (i) 2.42×10^{-12} m. (ii) 1.32×10^{-15} m.

T5 280 m.

T6 In units of eV, 0.98, 11.16, 13.05, 13.71, . . ., 14.56.

T7 (i) 35. (ii) An upper limit, since for $Z \gtrsim 35$ the orbit radius is less than the nuclear radius.

T8 Other possible wavelengths are 3.892×10^{-7} m, 7.500×10^{-7} m, 5.525×10^{-7} m, 1.067×10^{-6} m and 2.097×10^{-6} m. But, in fact, only 1.067×10^{-6} m occurs in practice, since quantum mechanical selection rules forbid the others.

T9 (i) Measuring in units of 10^{-19} J, four levels; (0) at 0.00 (arbitrary), (1) at 5.11, (2) at 6.83, and (3) at 9.95. A: (2)→(0). B: (1)→(0). C: (3)→(0), second order spectrum. D: (3)→(2). E: (2)→(1).
(ii) F: (3)→(1), second-order spectrum.
(iii) 11.80°.

T11 (i) $E_n = n^2h^2/8ma^2$. (ii) $x = \frac{1}{2}a$ for $n = 1$, $x = \frac{1}{4}a$ or $\frac{3}{4}a$ for $n = 2$.
(iii) $P = a^{-1}[\epsilon - (a/2\pi n) \sin(2\pi n\epsilon/a)]$. (iv) Equally likely to be found anywhere in $0 \leqslant x \leqslant a$, i.e. the classical result.

T12 $(h^2/8m)(a^{-2} + b^{-2})$.

T13 (ii) $E = hk^{1/2}/4\pi m^{1/2} = \frac{1}{2}h\nu$.

U Shorts

U1 No change.

U2 The copper is not attracted to either of the other two. Suspend each of the non-cuprous bars in the Earth's field to establish its direction (both will indicate the same); arrange them mutually orthogonally in a rigid cross and suspend this; the bar taking up its original position is the permanent magnet. Alternatively, hold the two bars in the form of a *T*. With the magnet as the 'upright' there will be a strong force pulling the two parts together; with the magnet as the 'crossbar' there will be little or no such force.

U3 Run a distance $d - wV_2(V_1^2 - V_2^2)^{-1/2}$ and swim the rest.

U4 $0.7\,\mathrm{m\,s^{-1}}$.

U5 1.

U6 $2.03\,\mathrm{kg\,m^2}$.

U7 ≈ 0.2 mA for a $100\,\mathrm{km\,h^{-1}}$ train on a 1.2 m gauge railway.

U8 $V(b-a)/2b$ upwards.

U9

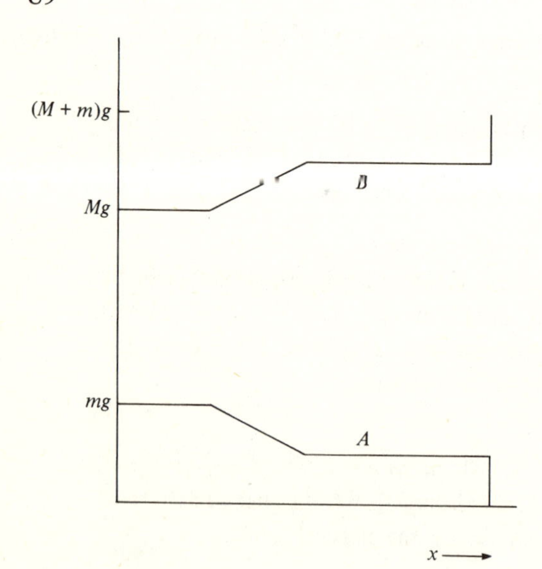

U10 $2 \times 10^{12}\,\mathrm{rev\,s^{-1}}$ if the nitrogen molecular diameter is taken as 10^{-10} m.

U11

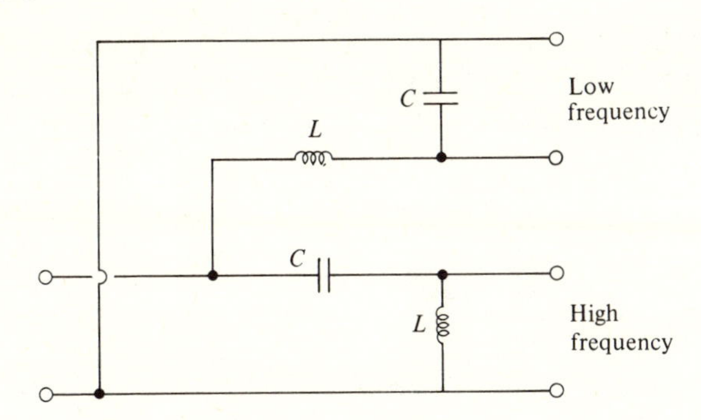

U12 The period depends upon the surface tension and the density, and upon the drop's radius to the $\frac{3}{2}$ power.

U13 The time needed is $\propto (\text{mass})^{2/3}$, assuming that the principal heat transfer process is conduction and the two eggs are made of similar material. If an ostrich egg weighs about 5 kg, compared with 50 g for a hen's egg, then about $1\frac{3}{4}$ hours is needed.

V Longs

V1 About 10^6.

V2 No. Each sub-pile has its centre of gravity above support from below. For example, the sub-pile consisting of the top three books has its centre of gravity $28\frac{1}{3}$ mm from the edge of the table, i.e. above the lowest book.

V3 (i)

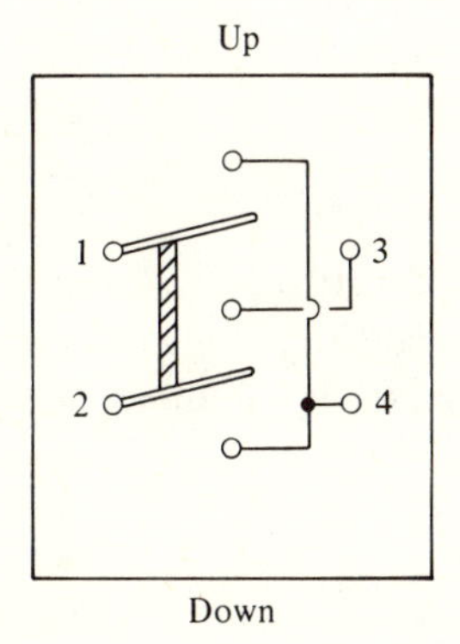

(ii)

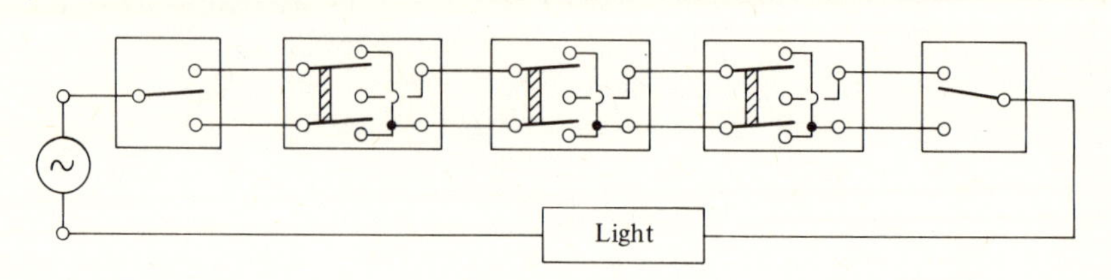

V4 $2h[(V_2+V_1)/(V_2-V_1)]^{1/2}$.

V5 32.5 ms.

V6 (i) The solid ball has a smaller value of k and reaches the bottom first. (ii) No. (iii) No, the analysis is independent of the mass and hence of the density.

V7 $2\pi[h\rho_I/3g(\rho_W-\rho_I)]^{1/2}=11.0\,s$.

V8 (i) $2(1+\epsilon_r)^{-1}$. (ii) The same as originally. The oil, having $\epsilon_r>1$, is attracted to where the electric field is strongest and work has to be done to remove it from there.

V9 The day gets longer.

V11 $d^{-1}(2mV/e)^{1/2}$.

V12 The patterns of current flow are as follows:

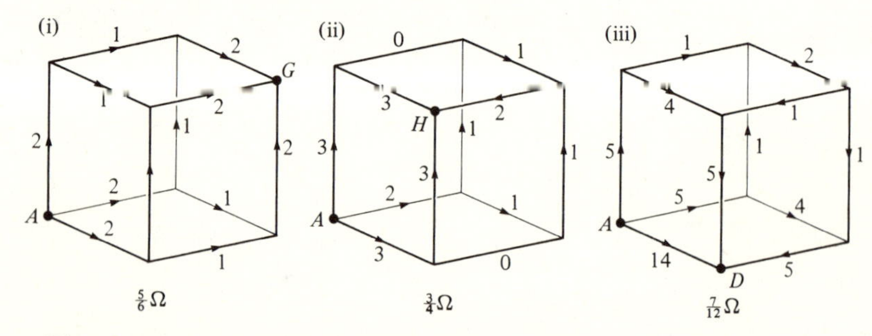

$\frac{5}{6}\Omega$ $\frac{3}{4}\Omega$ $\frac{7}{12}\Omega$

V13 The manoeuvre is possible provided the equation $[(6m/M)\sin\theta]+\theta=\pi$ has a solution θ less than π. Finding the condition for a maximum, and the value of that maximum, for $[(6m/M)\sin\theta]+\theta$, shows that it is necessary that $6m>M$.

W Data handling

W1 $9.80\pm0.02\,m\,s^{-2}$.

W2 307-316.

W3 (i) 55.1. (ii) Standard deviation of the mean ≈ 0.6. (iii) It is likely at the 3% level that the data are in conflict with the accepted value.

W4 207 m.

W5 (i) The data are in good agreement with the theory, except possibly at high input velocities where two successive points lie off the best fit line by about $1\frac{1}{2}$ standard deviations. (ii) $1.4 \times 10^9 \, \mathrm{N\,m^{-2}}$. (iii) The coefficient falls fairly rapidly from about 0.9 at negligible impact velocity to about 0.68 at an impact velocity of $0.9 \, \mathrm{m\,s^{-1}}$, and then more slowly as the impact velocity increases further.

W6 $a = 1.20, b = -3.29$.

W7 (i) 0.27. (ii) -0.26. (iii) 0.096. (iv) Estimate (ii) is the most accurate because, using the fact that $y(-x) = -y(x)$, it is effectively obtained by *inter*polation. Estimate (i) is close to the data region, but is also close to the point $x = (-\mathrm{b})^{1/2} = 1.8$ where y has a maximum, and so the accuracy of $y(1)$ is strongly dependent on that of both a and b. Estimate (iii) is further from the data region, but only needs an accurate value for a, since $12^2 \gg b$; a is quite well determined by the data. Thus, the order of likely accuracy is (ii) followed by (iii) and (i), probably in that order.

W8 (i) $y(T) = 3.03 \exp(-28.9/T)$. (ii) $y(5) \approx 9 \times 10^{-3}, y(500) \approx 2.86$. (iii) $a = 3.03, b = -88$.

W9 (i) Not in conflict. (ii) $X - Y(3.0 \pm 2.4)$ and $Y^2 - \pi^2 X(47 \pm 38)$ differ from zero (the theoretical value) by 1.2 standard deviations in each case. The two theories are supported equally by the data.

W10 (i) At the 1% level it is most unlikely that there is no correlation. (ii) There is no evidence for a claim that pupils either had a preference in both areas or none at all.

W11 The voting by the boys was almost certainly not random, but that by the girls could have been.

W12 (i) Yes. (ii) About 1 in 30. (iii) There is (only) about a 1 in 10 chance that he could have obtained his second set of results without the benefit of the instruction. (iv) He needs about $(0.360)^{-5} = 165$ attempts. (He was lucky to get a 5-gold group initially!)

W13 Overall probability of a fall = 0.118. There is only about a 1 in 30 chance that all fences are equally difficult. Fence 2 is much harder than average, whilst fences 4, 8 and 10 are rather easier.

X GuEstimation

X1 (*a*) 8×10^{21} for a bulb volume of $2 \times 10^{-7}\,m^3$.
(*b*) 1.3×10^4 kg for a lung capacity of 2 litres and a breathing rate of 10 times per minute.
(*c*) $2 \times 10^{-5}\,kg\,m\,s^{-1}$ for a 2 mm diameter drop falling at $5\,m\,s^{-1}$.

X2 (*a*) 1.6×10^{21} kg for an average depth of 4 km.
(*b*) Gains 45 s, if the coefficient of linear expansivity is $10^{-5}\,K^{-1}$ and the temperature drop is 15 K.
(*c*) 85 p at 4 p per kWh.

X3 (*a*) $530\,m\,s^{-1}$ at 300 K.
(*b*) A useful power of 20 kW for a cage and ten passengers together weighing 1000 kg moving at $2\,m\,s^{-1}$. With the use of counterweights this would be reduced.
(*c*) 10^{-5} kg for a bubble of thickness a few $\times 10^{-6}$ m, as indicated by the presence of interference colours.

X4 (*a*) 6 km for headlamps 1.2 m apart viewed by somebody with a pupil diameter of 3 mm, but see **H8**(*a*).
(*b*) $500\,m^3$ if the air in the balloon is 40 K hotter than that outside and the man has a mass of 70 kg.
(*c*) $1.2 \times 10^{-3}\,m^2$ if the absolute pressure in the tyre is 4 atmospheres, and the load equally distributed between the two tyres.

X5 (*a*) 1.2 kg for a car of mass 10^3 kg and an initial water temperature of 20 °C.
(*b*) 10^{26} for a 10% efficient 100 W bulb with a life time of 1000 hours.
(*c*) A turns ratio of 1.8 :1 would suffice in principle for a 0.6 mm gap and a breakdown field strength of $10^6\,V\,m^{-1}$.

X6 (*a*) 4×10^{-7} m.
(*b*) 1.4×10^5 K to lose the oxygen and nitrogen.
(*c*) At least 4 kV.

X7 (*a*) 100 N for a gut string of linear density $10^{-3}\,kg\,m^{-1}$, working length 0.35 m, and tuned to 440 Hz.
(*b*) 48 K treating the Earth as a uniform sphere of specific heat capacity $800\,J\,kg^{-1}\,K^{-1}$ and the heat as uniformly distributed.
(*c*) 150 mm for a bowl of radius 100 mm.

X8 (*a*) 0.2° for a piece of a record which plays for 20 minutes whilst tracking inwards by 100 mm.
(*b*) 2.3×10^{12} J for a pyramid of base length 230 m, height 147 m and made of stone of density 2.5×10^{3} kg m^{-3}.
(*c*) 7.5 K assuming the Earth radiates as a black-body.

X9 (*a*) 7 W for a 10 g grasshopper which can jump to a height of 0.3 m as a result of straightening its legs and thereby raising its centre of gravity by 5 mm whilst still in contact with the ground.
(*b*) 4×10^{7} s^{-1} for a nitrogen molecule of assumed cross-sectional area $\pi \times 10^{-20}$ m^2 at 300 K.
(*c*) 1.6 mJ for a bar 200 mm long and cross-sectional area 10^{-4} m^2 with 600 turns m^{-1} of conductor.

X10 (*a*) 7×10^{6} m^2 if the conversion inefficiency and atmospheric absorption produce a loss of a factor of 10.
(*b*) 2.7 kW for a house of volume 600 m^3 and an inside–outside temperature difference of 17 K, assuming a molecular specific heat of $\frac{5}{2}k$ for nitrogen/oxygen at room temperature.
(*c*) 77 m taking the density of copper as 8.9×10^{3} kg m^{-3} and its resistivity as 1.8×10^{-8} Ω m.

X11 (*a*) 10^{-8} W if the volume available is 6×10^{-7} m^3 and the (one-way) variation in the atmospheric pressure in a week is 10%.
(*b*) 2.4 mK, taking the surface tension of water as 72 mN m^{-1}.
(*c*) 2 MJ for a cloud of area 10^{7} m^2, 500 m above the Earth's surface.

X12 (*a*) 6×10^{3} m s^{-2} for a 20 minute record which tracks across 100 mm, assuming that the recording amplitude is one-quarter of the groove separation and that notes of frequencies up to 2 kHz are recorded at this amplitude.
(*b*) About 2×10^{-5} assuming that the average ocean depth is 4 km and that the average vapour pressure of water in the atmosphere is 5 mm of mercury.
(*c*) 17 batteries (12 V, 20 A h) if a car engine is about 25% efficient and the car will run for 20 minutes at 50 km h^{-1} on one litre of petrol.

X13 (*a*) 4.3 Hz if the cross-sectional area of the band does not change too much on stretching.
(*b*) 3800 years taking the density of coal as 1.4×10^{3} kg m^{-3}.
(*c*) 2×10^{4} assuming complete mixing of the Earth's water molecules, that the average depth of the oceans is 4 km, and that the volume of a glass is 0.5 litres.

APPENDIX: POACHER TURNED GAMEKEEPER

Just as few students will admit to finding pleasure in the taking of examinations, most examiners view examination marking as a job to be done with care rather than delight. However, from time to time, questions are set, the offered answers to which provide for the examiner some unexpected entertainment.

Presented below is a small selection of such answers and partial answers that have been offered to the author in the past. They are reoffered to the reader for both instruction and, it is hoped, mild assessment. The reader is invited to say what, if any, merit they see in the proffered answers, comments or assumptions.

Q. It has been said that every breath you take contains several molecules from the dying breath of Julius Caesar. Verify the basis for this statement.

Answers

1. 'Italy is fairly warm so let us suppose that the temperature was 27 °C even though it was March . . .'
2. 'Assume Caesar was a normal man . . .'
3. 'Assume Caesar's lungs were a little on the small side $\approx$2.24 litres.'
4. 'Assume J.C.'s lungs were 10 cc . . .'
5. 'Assume J.C.'s dying breath contained approximately 10 moles of air . . .'
6. 'There are about 10^{11} people breathing on this earth every minute . . .'
7. 'Julius Caesar must have taken about two minutes to die; during which . . .'
8. 'The distance from Rome $\approx$900 miles . . .'
9. 'Romans were not large people and the lung capacity of Caesar would be . . .'
10. '. . . the temperature at the Capitol being greater than 0 °C, . . .'
11. 'Thus as Caesar died those surrounding him must have absorbed some of the CO_2 . . . which has been passed on from mouth to mouth.'
12. 'Assume he died at N.T.P.'
13. 'Since the lungs are open to the air we may assume the pressure is atmospheric . . .'
14. '. . . it is evident that at least one molecule must be breathing. This follows from Dalton's Law . . .' (*sic*)

15. 'Julius Caesar was assassinated, say at the age of 40, ...'
16. 'Volume of air in dying breath $= 50 \times \pi 2^2$ c.c. ...'
17. 'Assume volume of Julius Caesar = 50 litres ...'
18. 'Since Julius Caesar is composed mostly of water ...'
19. 'Now there are approximately 2×10^9 people in the world, and this is more than ever before, yet ...'
20. 'J.C.'s air intake would be very small if he was dying ...'
21. 'The cycle may have been started by someone breathing in near Caesar ...'
22. 'Let volume of air breathed be $V\,\text{cm}^3$ (assume the same for Caesar as for mortals) ...'

Q. A person walks briskly down a road beside a fence composed of vertical posts 0.2 m apart. Explain carefully what he hears as the echo of the tapping of his shoes on the road.

Answers

1. 'Half the taps will pass through the fence.'
2. 'Refraction is occurring since the gap between the posts is of the same order of magnitude as the sound of the footsteps.'
3. 'The footsteps set the posts swinging with their resonant frequency and causes damage - driven by the pavement.'

Q. An athlete can just jump over a bar 2 m above the ground. What mean force does he exert on the ground just before take-off?

Answers

1. 'Let the mass of the athlete be 2 kg.'
2. 'Take the athlete's mass as 330 kg.'
3. 'The force is 800.03104 N.'
4. 'Length of leg of athlete = 3 m.'

Q. Discuss the following statement: The dimensions of a regular block are measured with a metre rule to be 0.512 m × 0.126 m × 0.373 m. The volume of the block is therefore $0.024062976\,\text{m}^3$.

Answers

1. 'The answer given is incorrect, but it is very easy to make a slip with a decimal point. The correct answer is 0.24062976.'
2. 'Molecules are separated by spaces which account for the difference between .024062976 and .02406 (calculated by logs).'
3. '$V = \int_0^l \int_0^b \int_0^h \mathrm{d}x\, \mathrm{d}y\, \mathrm{d}z = lbh$.'

4. 'A cubic metre is a much larger unit than a metre is length.'
5. 'The stated volume is to the correct number of places but the actual number is wrong = .004709476 m^3.'
6. 'The ratio of 1 metre to 1 cubic metre is 1:3.'
7. 'The volume appears to be less than even the height, but this is only the numerical value. When three quantities less than one are multiplied, the product is always much smaller. The surface area, however, is much larger.'
8. 'For convenience and neatness it is better to express it in cm^3 so that the volume becomes 24062.976 cm^3.'
9. '... but due to human error on multiplication the volume cannot be calculated to 9 places of decimals.'
10. 'This is because the act of measurement causes a slight force to be exerted on the material. This force will squeeze the molecules closer together and the measurement will not be correct, but too small.'

SYMBOLS AND UNITS

Symbol		Quantity	Unit
A		(cross-sectional) area	m^2
		displacement amplitude	m
		mass number (nuclear number)	
		refracting angle of prism	rad
		abundance	
a		length, width, radius	m
		displacement amplitude	m
		acceleration	$m\ s^{-2}$
B		magnetic flux density	T
		displacement amplitude	m
		binding energy of a nucleus	J
b		depth, width, radius	m
C		capacitance	F
		heat capacity	$J\ K^{-1}$
		couple	N m
		binding energy per nucleon in a nucleus	J per nucleon
	C_v	molar heat capacity at constant volume	$J\ mol^{-1}\ K^{-1}$
	C_p	molar heat capacity at constant pressure	$J\ mol^{-1}\ K^{-1}$
c		speed of light in a vacuum	$m\ s^{-1}$
		specific heat capacity	$J\ kg^{-1}\ K^{-1}$
	$\bar{c}$	mean speed	$m\ s^{-1}$
	$\overline{c^2}$	mean square velocity	$m^2\ s^{-2}$
	c_v	specific heat capacity at constant volume	$J\ kg^{-1}\ K^{-1}$
	c_p	specific heat capacity at constant pressure	$J\ kg^{-1}\ K^{-1}$
	$c_{v,m}$	molar heat capacity at constant volume	$J\ mol^{-1}\ K^{-1}$
	$c_{p,m}$	molar heat capacity at constant pressure	$J\ mol^{-1}\ K^{-1}$
D		distance	m
		angle of (minimum refracted) deviation	rad

d		distance, spacing, diameter	m
E		energy, kinetic energy, total energy	J
		electric field strength	$V\ m^{-1}$
		Young modulus	$N\ m^{-2}$ (Pa)
		electromotive force	V
	$\mathcal{E}$	electromotive force, ionization potential	V
	E_0	breakdown electric field strength	$V\ m^{-1}$
e		elementary charge	C
		strain	
	e	base of natural logarithms	
F		force, tension	N
		force per unit length	$N\ m^{-1}$
f		force	N
		frequency	Hz (s^{-1})
		focal length	m
		arbitrary function	
G		gravitational constant	$N\ m^2\ kg^{-2}$
g		acceleration of free fall	$m\ s^{-2}$
		arbitrary function	
H		heat production rate	$W\ m^{-3}$
h		Planck constant	J s
		height	m
		head of liquid	m
I		current	A
		moment of inertia	$kg\ m^2$
		light intensity	$W\ m^{-2}$
i		current	A
J		angular momentum	$kg\ m^2\ s^{-1}$
k		Boltzmann constant	$J\ K^{-1}$
		spring-constant	$N\ m^{-1}$
		torsion constant	$N\ m\ rad^{-1}$
		wave number	m^{-1}
		radius of gyration	m
L		length	m
		inductance	H

l	length	m
	specific latent heat	$J\ kg^{-1}$
M	magnetic moment	$J\ T^{-1}$
	mass	kg
M_S	mass of Sun, star, etc.	kg
m	integer	
	mass, mass of molecule	kg
$m_{e;p;n;\pi}$	mass of electron; proton; neutron; π-meson	kg
m_0	rest mass	kg
N	integer	
	neutron number	
N_A	Avogadro constant	mol^{-1}
n	integer	
	number of turns	
	order of spectrum	
	label of energy level	
	number of moles of gas	
	number density	m^{-3}
	number of turns per unit length	m^{-1}
	refractive index	
$n_{G;W}$	refractive index of glass; water	
P	power	W
	pressure, pressure difference	$N\ m^{-2}$ (Pa)
	impulse	N s
	probability	
p	pressure, pressure difference	$N\ m^{-2}$ (Pa)
	momentum	$kg\ m\ s^{-1}$
	path difference	m
	probability of success in binomial trials	
Q	electric charge	C
	quantity of heat	J
	energy release, Q value	J
R	resistance	Ω
	gas constant	$J\ mol^{-1}\ K^{-1}$
	normal reaction	N
	radius, distance	m
$R_{E;S;M}$	radius of Earth; Sun; Moon	m

$R_{ES;EM}$	Earth–Sun; Earth–Moon distance	m
r	radius	m
	resistance	Ω
S	energy flux	$W\ m^{-2}$
	entropy	$J\ K^{-1}$
s	distance	m
T	absolute (thermodynamic) temperature	K
	kinetic energy	J
	periodic time	s
	tension	N
t	time	s
	thickness	m
U	speed, velocity	$m\ s^{-1}$
u	speed, velocity	$m\ s^{-1}$
	object–lens distance	m
V	volume	m^3
	potential energy	J
	voltage, potential difference	V
	speed, velocity	$m\ s^{-1}$
V_E	speed of Earth in orbit	$m\ s^{-1}$
v	volume	m^3
	image–lens distance	m
	speed, velocity	$m\ s^{-1}$
v_d	drift speed	$m\ s^{-1}$
W	energy, work	J
	number of arrangements of a thermodynamic system	
w	width	m
x, y, z	position, distance, displacement	m
Z	atomic number (proton number)	
α	angle	rad
	linear expansivity	K^{-1}
	reflection coefficient	
	fraction	

Symbol		Quantity	Unit
β		angle	rad
γ		surface tension	$N\ m^{-1}$
		ratio of principal specific heats	
ϵ		phase angle	rad
		strain	
		binding energy	J
	ϵ_0	permittivity of a vacuum	$F\ m^{-1}$
	ϵ_r	relative permittivity	
η		viscosity	$kg\ m^{-1}\ s^{-1}$ (Pa s)
θ		angle	rad
		common temperature	°C
λ		wavelength	m
		thermal conductivity	$W\ m^{-1}\ K^{-1}$
		nuclear decay constant	s^{-1}
μ		mass	kg
		mass per unit length	$kg\ m^{-1}$
		coefficient of friction	
	μ_0	permeability of a vacuum	$H\ m^{-1}$
ν		frequency	Hz (s^{-1})
π		pi-meson	
		3.141 ...	
ρ		density	$kg\ m^{-3}$
		resistivity	$\Omega\ m$
		resistance per unit length	$\Omega\ m^{-1}$
σ		Stefan–Boltzmann constant	$W\ m^{-2}\ K^{-4}$
		stress	$N\ m^{-2}$ (Pa)
		surface tension	$N\ m^{-1}$
		electrical conductivity	$S\ m^{-1}$ ($\Omega^{-1}\ m^{-1}$)
		surface charge density	$C\ m^{-2}$
		linear charge density	$C\ m^{-1}$
τ		time interval, time constant	s
	$\tau_{1/2}$	half-life	s
Φ		magnetic flux	Wb
		cumulative Normal distribution function	

ϕ		angle	rad
		phase angle, phase difference	rad
		pressure gradient	$N\,m^{-3}$ ($Pa\,m^{-1}$)
		heat flux	$W\,m^{-2}$
	ϕ_0	work function	J
χ	χ^2	chi-squared statistic	
ψ		quantum wave function in *n* dimensions	$m^{-n/2}$
ω		angular speed, angular velocity, angular frequency	$rad\,s^{-1}$

FORMULAE AND RELATIONSHIPS

Mechanics

Uniform acceleration: $v = u + at; s = ut + \frac{1}{2}at^2; v^2 = u^2 + 2as$
Force: $f = \mathrm{d}p/\mathrm{d}t; f = ma$
Work: $W = fs$ (s in direction of f)
Momentum: $p = mv$
Kinetic energy: $T = \frac{1}{2}mv^2$
Circular motion: $v = \omega r; a = v^2/r = \omega^2 r$
Rotational motion: $J = I\omega; T = \frac{1}{2}I\omega^2; C = I(\mathrm{d}\omega/\mathrm{d}t)$
Simple harmonic motion:

$x = A \sin(\omega t + \epsilon) = A \sin(2\pi ft + \epsilon)$
$a = -\omega^2 x; T = 2\pi/\omega; E = \frac{1}{2}m\omega^2 a^2$
$\omega^2 = g/l$ (simple pendulum)
$\omega^2 = k/m$ (mass on spring)

Elastic spring: $f = kx; E = \frac{1}{2}kx^2$
Elastic wire: $\sigma = E\epsilon; W = \frac{1}{2}\sigma\epsilon = \frac{1}{2}E\epsilon^2 = \sigma^2/2E$
Gravitation:

$f = Gm_1m_2/r^2; V = -Gm_1m_2/r$
$V = mgh$ (effectively uniform field)

Friction: $F = \mu R$
Flotation: $mg = f = V\rho_L g$

Waves and light

Speed:

$v = f\lambda$ (general); $c = f\lambda$ (electromagnetic waves)
$c = (\epsilon_0\mu_0)^{-1/2}$ (electromagnetic in a vacuum)
$v = (F/\mu)^{1/2}$ (transverse on string)
$v = (E/\rho)^{1/2}$ (sound in solid rod)
$v = (\gamma p/\rho)^{1/2}$ (sound in gas)

Doppler: $f'/f = (v - u_0)/(v - u_s)$ (v, u_0, u_s all measured in same direction)
Refraction: $\sin\alpha_I = n \sin\alpha_R$

Prism: $\sin \frac{1}{2}(A + D) = n \sin \frac{1}{2}A$ (minimum deviation)
Lens: $u^{-1} + v^{-1} = f^{-1}$; magnification $= v/u$
Diffraction:

$d \sin \theta = n\lambda$
$y = n\lambda D/d$ (double slit)
$\Delta\lambda/\lambda = 1/nN$ (grating resolving power)
$\sin \theta = \lambda/d$ (single slit, first minimum)
$\sin \theta = 1.22\lambda/d$ (circular hole, first minimum)

Bragg reflection: $2d \sin \theta = n\lambda$

Heat and kinetic theory

Heat capacity: $\Delta Q = C\Delta T$
Ideal gas:

$pV = nRT$ (for n moles)
$p = \frac{1}{3}nmc^2$ (for number density n)
$N = \frac{1}{4}n\bar{c}$ (number of molecules striking unit area of the walls in unit time)
$E \approx \frac{3}{2}kT$ (thermal translational kinetic energy of a molecule)
$C_p - C_v = R$
$C_p/C_v = \gamma$
$pV =$ constant (isothermal change)
$pV^\gamma =$ constant (adiabatic change)

Thermal conduction: $\phi = \lambda\ (\mathrm{d}T/\mathrm{d}x)$
Black body: $P = \sigma T^4 A$
Entropy: $\Delta S = k \ln (W_2/W_1)$

Electricity

Point charges: $f = Q_1Q_2/4\pi\epsilon_0 r^2$; $V = Q_1Q_2/4\pi\epsilon_0 r$; $E = Q/4\pi\epsilon_0 r^2$
Electric fields: $E = -\mathrm{d}V/\mathrm{d}x$; $E = V/d$ (uniform field)
Capacitors:

$Q = CV$; $E = \frac{1}{2}QV = \frac{1}{2}CV^2 = Q^2/2C$
$C = C_1 + C_2 + \ldots$ (parallel); $C^{-1} = C_1^{-1} + C_2^{-1} + \ldots$ (series)
$C = \epsilon_r\epsilon_0 A/d$ (parallel plate); $C = 4\pi\epsilon_0 a$ (isolated sphere)

Direct current: $V = RI$; $P = IV$; $P = RI^2$; $I = neAv_d$
Resistors:

$R = R_1 + R_2 + \ldots$ (series); $R^{-1} = R_1^{-1} + R_2^{-1} + \ldots$ (parallel)
$R = \rho L/A$ (uniform conductor)

RC circuit: $\tau = RC; Q = Q_0 \exp(-t/\tau)$
RL circuit: $\tau = L/R; I = I_0 \exp(-t/\tau)$
LC circuit: $2\pi f = \omega = (LC)^{-1/2}$
Reactance: $(\omega C)^{-1}$ (capacitor); ωL (inductor)

Electromagnetism

Force on moving charges:

$f = BQv \sin\theta$ (charge); $f = BIL \sin\theta$ (current-carrying wire)

Fields due to currents:

$\int B \, \mathrm{d}l = \mu_0 I$ (around a closed path)
$B = \mu_0 I/2\pi r$ (long straight wire)
$B = \mu_0 NI/2a$ (centre of circular coil)
$B = \mu_0 nI$ (long solenoid)

Electromagnetic induction:

$\mathcal{E} = -\mathrm{d}\Phi/\mathrm{d}t; \Phi = LI$
$V = L(\mathrm{d}I/\mathrm{d}t)$ (voltage needed to sustain rate of change of current)

Modern physics

Nuclear decay: $\mathrm{d}N/\mathrm{d}t = -\lambda N; N = N_0 \exp(-\lambda t); \tau_{1/2} = 0.693/\lambda$
Photons: $E = hf; p = hf/c = h/\lambda; T_{\mathrm{max}} = hf - \phi_0$ (photoelectric effect)
Particles: $\lambda = h/p = h/mv; E = m_0 c^2$
Nuclear reaction: $c^2 \sum m_{\mathrm{initial}} = c^2 \sum m_{\mathrm{final}} + Q$

CONSTANTS

Speed of light in a vacuum, $c = 3.0 \times 10^{8}$ m s^{-1}.

Elementary charge, $e = 1.60 \times 10^{-19}$ C.

Mass of electron, $m_e = 9.1 \times 10^{-31}$ kg.

Mass of proton, $m_p = 1.67 \times 10^{-27}$ kg.

Avogadro constant, $N_A = 6.0 \times 10^{23}$ mol^{-1}.

Planck constant, $h = 6.6 \times 10^{-34}$ J s.

Boltzmann constant, $k = 1.38 \times 10^{-23}$ J K^{-1}.

Stefan-Boltzmann constant, $\sigma = 5.7 \times 10^{-8}$ W m^{-2} K^{-4}.

Gravitational constant, $G = 6.7 \times 10^{-11}$ N kg^{-2} m^{2}.

Gravitational acceleration, $g = 9.8$ m s^{-2}.

Permeability of a vacuum, $\mu_0 = 4\pi \times 10^{-7}$ H m^{-1}.

Permittivity of a vacuum, $\epsilon_0 = 8.8 \times 10^{-12}$ F m^{-1}.